Quality Management Integration: Quality In Modern Business Abridged Edition

Warren Alford

Quality Management Integration

© 2011 Warren Alford

www.WarrenAlford.com

Published by
Speckbohne Publishing
All rights reserved

Abridged - First Edition

10 9 8 7 6 5 4 3 2 1

Quality Management Integration

Dedicated to
Silvia and Lola
I Love You...

Quality Management Integration

Quality Management Integration

Table of Contents

Quality Management Integration

Introduction

Quality - *A subjective term for which each person or sector has its own definition. In technical usage, quality can have two meanings:* **1.** *The characteristics of a product or service that bear on its ability to satisfy stated or implied needs;* **2.** *a product or service free of deficiencies.*[1]

Quality is a word, a theory, an idea, a system, an assumption and a sales tool. It can convey different thoughts to each person's mind. Certainly, our individual definitions are as varied as opinions of style, fashion and taste. It is a small but important word, which bears more importance than most people realize.

It is interesting to note most people today cannot truly define quality when speaking of a management philosophy or a measurement tool. Quality is simply a *Sales* tagline or buzzword that is used to lure potential customers to the front door of a business.

I submit this question; *"What does Quality truly mean to you?"* Before you answer the question, I believe your response will probably be in line with your professional position.

A *CEO* may believe quality gives his / her company an edge over the competition. This may mean less customer complaints and greater customer satisfaction.

To some *managers*, it may simply mean once a month they must deal with *pesky* auditors and inspectors who simply rip through their departments and tell them what is being done wrong. The *Sales Department* may use the word *quality* to close a deal, win a contract or attract other customers.

The *customer* will surely have his or her own idea of *quality*, which may equate to not having to return a defective purchase or hassle with very much *"service after the sale"*. Perhaps it is the sole reason a customer is willing to pay *more for your* product than your competitor's product. Being able to rely on receiving the same *or* a higher standard product or service during a purchase may obviously be a deciding factor.

So, how do you begin to develop *quality*? How do you create a system that customers will believe in and grow to trust? Is it enough to simply use the word *quality* in your advertising? Perhaps changing your web site and including the word quality more frequently will result in quality being seen through the customer's eyes. Is it enough to ask employees to use the word quality as a new buzzword? Is *quality* something more elusive?

You may think of quality in this manner and you will quickly realize quality is not a coincidence or happenstance. Quality is the result of a long, difficult, sometimes confusing and resource-draining exercise that only true upper management skills can put into place and see to fruition.

Quality is not an accident. It requires money, time and other resources to put into place. It is a system that will, one day in the future, allow you to harvest a premium crop that will **bring the customers to you**.

When the economy is good, anyone can manage a business and have a favorable chance of earning a profit. The *talented* leaders stand out when the economy is bad and profits are hard to find. It is the foresight and a keen ability to believe in quality in the midst of company layoffs and downsizing. The system of quality requires the belief in the system and the knowledge that quality *can and does* contribute greatly to the company bottom line.

The reasoning behind this statement is simple: when the economy is bad, your customer has less money to spend. If the customer can consistently depend on your product and / or service, they are able to better budget *their* business.

Your customer may try your competitors but if their poor quality results in unexpected expenses (i. e. rework, product failure for their customers, etc.) they will not only come back to your company but will actually pay a higher price to ensure your product / service is always *available to them* in the future.

If your customer pays $10.00 less to your competitors and then spends $30.00 to fix the problems created, they would probably pay you $30.00 more to ensure no problems occur in the first place. Also remember, it is difficult to quantify *customer loyalty,* which may cost considerably more after damaging the business relationship with the customer.

Quality philosophies are also as varied as customers themselves. The common denominator is the perception the customer has of your *approach to quality.* If they know you are consistent in your application and you strive for continuous improvement and excellence, they are more likely to be understanding *if and when* a defect does occur.

As the saying goes, *"Actions speak louder than words".* This truly applies when it comes to quality. If you strive for the *minimum,* your customers will know and may not be *your* customers much longer.

An organization that takes quality seriously in approach and application will begin to undergo a metamorphosis that will even surprise the management who put the system into place. Quality is truly a living presence that will take on a life of its own.

The quality principle must be ingrained from upper management, all the way *down* through the organization. Without this simple, yet critical element, there will never be a true quality system and you are paying the Quality Department to *"pretend and play quality"*. Management who has seen quality truly working rarely recedes into the *"old way of thinking"*. After seeing what quality can do, it is impossible for most to revert back to *quality lip service.*

Change is a constant that many people fear. For a quality system to exist, there must first be change. The way a person thinks and rationalizes situations must change when forming the quality base.

One must understand that failure is as viable a possibility as success. Either way, the organization will never be the same. It will either grow and flourish in the quality model or it will soon fail in the endeavor; probably destroying all its customer relationships during the failure.

Yet, some members of upper management are adamant in reinforcing quality and the continuous improvement initiative. They have experienced quality first hand and know it makes their job easier, more predictable and ultimately more profitable.

Upper management who uses *Management By Financial Objective* will value their yearly bonuses more that quality but without quality, their bonuses will end as they drive customers away. No amount of reassurance matters when you can't fulfill your customers' requirements. Without quality, you *can not* and *will not* meet the customer's expectations and specifications with any degree of accuracy and consistency.

Other executives use quality to *"hedge their bets"* in a metaphorical way. In a world of changing business variables, quality is a constant that management can rely upon. No matter what the business or economic environment, the output of their processes remain strong, predictable and reliable. At the end of the day it is really about meeting the needs of the customer and without quality, you are relying upon luck instead of science. You will, statistically speaking, have a better chance of winning the lottery.

We will examine this phenomenon called *Quality* and explore the theory and application of the elements required to ensure success. It is within reason to expect some of the things presented herein to be dismissed by advocates of the *"fly by the seat of their pants"* management theory. These are usually the managers who destroyed the moral of their subordinates with distrust, bullying and deception. If the energy were better used for true management and quality integration measures, they would find a plethora of success in their business endeavors.

In the following pages you will find a roadmap to assist in leading your organization to success. We will begin with the basics, which some may find a waste of time *however* a building is constructed on a foundation or else it falls.

Quality is no different. If you skim over the basics, you will certainly find it difficult to understand the theory in the following pages. Many mangers claim to understand the human element but in reality, very few do. It is this human element that will prove to be the greatest challenge in creating a quality system. Throughout history, many armies have turned on their leaders and the *corporate workforce is no different*. The battle is won by the soldier, not the commander.

You cannot manage an organization without employees and you must truly be a student of psychology to understand the unpredictable variable that is the *employee*. Your management philosophy aside, if you misjudge and do not understand their needs, wants and motivations you will be setting yourself up for failure.

The basics of quality will lead to more advanced techniques and we will explore some of these areas as well. We are delving into a radical management theory and some of the material may seem a waste of time and energy. I remind you - *it is not.*

I urge you to consider *all* of the information and how it fits together in a puzzle - like manner. One element omitted during implementation and integration may have an adverse effect on the outcome of your quality pursuit.

"The significant problems we face cannot be solved at the same level of thinking we were at when we created them." Albert Einstein

Chapter 1

The theory of business is fairly simple. You need to create a product or service, find someone who wants what you created and convince them to purchase the product / service *you* are selling. It is easy to *explain*, easy to *understand* and extremely hard to *do*. If it were easy, everyone would do it.

All businesses face this challenge. The elements of business are basically the same whether you are selling widgets or a variety of services that someone would see as a value to them. Money is spent to create, market, sell, deliver and guarantee the product or service.

It is surprising that quality is an after thought for some companies who believe a *Quality Inspector* at the end of an assembly line is *all* the quality management needed.

What is interesting to point out is the fact that once the product reaches the end of the assembly line, it is already complete and no amount of inspection will be able to fix the defects. By this time, the damage is already done.

The idea of adding quality into the last few process steps is really a holdover from the days when farmers planted seeds, tilled the ground and at a certain time, gathered the crop.

After reaping the harvest, the farmer would use a rudimentary quality process resembling modern day quality control. The need to *"cull and grade"* the harvest led to the *"make it, then sort it"* approach. This technique is still used in many inspection - based quality system initiatives.

There is a basic flaw in using this approach exclusively; quality will be an afterthought and not an integral part of creating the product or service. If you use this quality theory in your business, you must be prepared to *"cull"* the defectives, which cost as much to make as non-defectives but carry the *additional* cost of rework, lost materials, wasted resources and scrap costs. Just like the farmer, you may end up throwing away a large percentage of your output.

Craft workers were the first to realize they had some control over their processes and they began to take advantage of that knowledge. The manufacturing plant would make the same product day in and day out. The problem was *consistency*.

The output from the assembly line was so varied there was no way to say with any certainty that a specified percentage of the output would meet the customer's specifications and standards.

This created a major problem for the manufacturer…especially when the customer imposed monetary sanctions on the manufacturer due to contractual nonperformance. If the contact called for one hundred defect-free widgets produced per day and the manufacturer could only provide seventy-five per day, the manufacturer would be required to drop their price per unit or pay a penalty. At this point, quality began to be used to help determine *why* the output consistency fluctuated. Quality was then tasked with discovering the *root cause* and evaluating a corrective action. Quality was about to enter the business world and become a tool that could not only identify problems but would seek ways to make the operation run smoother, faster and more economical. This could never be achieved with only a *"Quality Inspector"*.

The idea of manufacturing huge numbers of products and sorting out the good ones was an expensive undertaking. The factories of the early 1900s began to see that money could be saved and more could be earned by lowering the number of defectives and the way to do that was simple yet complicated at the same time. Quality would need to be incorporated into the entire manufacturing process; from receiving the raw materials, to loading the product into a truck or train for final transport. All steps in between would require quality controls.

There would need to be a monitoring system, which could objectively verify the processes were working and the output from each process was consistent. The idea was if all processes are in control and outputs are consistent, the widget at the end of the assembly line would be of consistent quality because there would be no opportunity for a defect to occur.

Manufacturers became obsessed with consistency but the harder they tried, the more they failed. But how could the numerous in - control processes not yield a product free of any defects?

Enter *variation*.

Variation meant even while a process was in control and providing constant output, the process was still producing variations. The process was in control, but being in control and being variation - free are two separate things. If each process varied 2% and there were twenty processes needed to manufacture a product, you could say the amplified sum of the output was *40%*. If your chances of waking up tomorrow are *40%*, you might want to increase your life insurance coverage before going to bed. If the process output varied so much, there was no way to reasonably expect to lower the occurrence of defects.

Great strides in process control were made during World War II. The military needed huge amounts of materials and uniform products. If a grenade fuse was supposed to take five seconds to detonate and it detonated in only two seconds, there was more at stake than *customer satisfaction*.

Quality Management Integration

Statistical Process Control (SPC)^2 was used extensively during the war by engineers. Reliability Engineers focused on designs and practices to build quality into the product from the very beginning. Elementary plotting charts gave way to more advanced statistical metric charts and the theory of "*management by facts and data*" was created.

Management was beginning to see that processes were "*alive*" in a certain sense and as such, could not be fully controlled and the variation completely removed. The challenge was to lower the variation to a point so it would no longer have an adverse effect on the process output. Since this was almost like foretelling the future, management again turned to the engineers for quality help.

Mathematics could *almost* foretell the future - at least a probability of occurrences. Predictions could be made by making assumptions of things which remained relatively stable and the *Law of Averages* could be used to gage certain outcomes. A process output is actually an *outcome*, so the principle could be applied.

Equally difficult was the management of the human element. Variation was *always* present when human interaction with the processes took place. The variation was totally unpredictable and huge fluctuations were always observed.

It was interesting to note that process output performance increased and variation decreased slightly in the short period of time prior to the employee's *payday*.

The human element remains the most unpredictable force in the modern organization. Human resources void *SPC* charts when human interaction is present. Employees also represent one of the highest (*if not the highest*) risk to an organization. Payroll is often over fifty percent of an organization's expense. Therefore, if we cannot predict or unequivocally control employees, there is only one option left. We must learn to "*read and understand*" the employee; their limitations, expectations and most of all, their motivations.

Without understanding the employee, the business will suffer greatly. Recognizing an employee can be an *asset or a liability* is one of the first steps the manager takes in becoming a *leader*.

Chapter 2

There are many things that motivate people. The difficulty faced by management is discovering *what* motivates the workforce. This is an age-old dilemma, which very few have mastered.

Some managers believe the workforce is motivated only by monetary means however, that is far from the truth. An employee must feel they are adequately and competitively compensated for their work, and they must earn enough money to pay their pecuniary obligations but here are many other variables to consider.

Managers should review the chart below (*Figure 1*). The ranking of work factors is quite surprising.

Factor	Employee Rating	Manager Rating
Interesting Work	1	5
Appreciation	2	8
Involvement	3	10
Job Security	4	2
Compensation	5	1
Promotion	6	3
Working Conditions	7	4
Loyalty to Employees	8	7

Help with Problems	9	9
Tactful Discipline	10	6

Figure 1: Ranking of Work Factors

A manager must align their actions with the employee's expectations. The manager who believes compensation is the most important factor may be simply *"throwing money at the problem"*.

This assumption can be very costly and may actually prove to be a detriment to the business. There must be more thought dedicated to how the manager intends to direct the workers to accomplish their goals. Money alone will not motivate the employee to give *110%* to the manager's agenda. So what can the manager do to further *their* plan?

Abraham Maslow's *Hierarchy of Needs (Figure 2)* theory considers human motivation to be based upon a simple premise: Individuals are motivated toward lower-order needs until they are satisfied. Once satisfied, the individual is free to pursue higher-order needs until ultimately, their *self-actualization* needs are fulfilled. His idea was management should place more emphasis on the higher-order needs for employees.

Self-actualization allows the employee to achieve their full potential and creativity, independence and spontaneity. Individuals who operate in this cognitive *"higher state"* are the ones who can get things done the right way and in a timely manner. Alignment of their goals with the goals of the manager can be a win-win situation for manager, employee and company.

Observant managers can use this system to their benefit by adjusting the employee's perception; showing the employee what is in it for them to align with the manager's goals.

Once the goals are made common, the manager has not only directed the employee to the destination but has also motivated the employee to strive and work harder to achieve the objective.

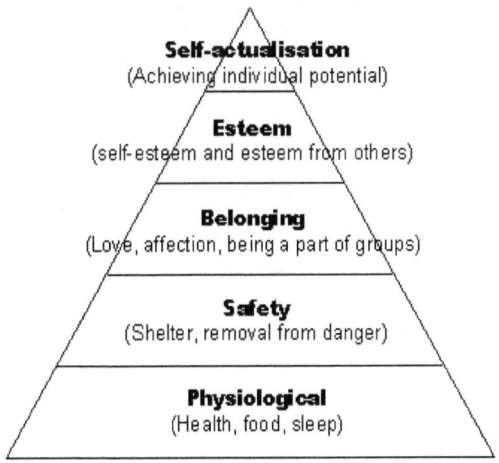

Figure 2: Maslow's Hierarchy of Needs

Quality Management Integration

Theory X and *Theory Y (Figure 3)[3]* are two dominant theories in industrial practice. They are readily recognized by most managers in the modern workplace. These theories are based on how managers view their employees and how they are treated for motivational and work related purposes.

Theory X	Theory Y
Dislikes Work	Willing to Work
Lacks Ambition	Responsible
Irresponsible	Self Directed
Will Not Change	Has Self Control
Follower	Leader

Figure 3: Comparison of Theory X and Theory Y

Douglas McGregor viewed management's job as one in which working conditions are created so that individuals can establish and integrate goals with those organizations.

The basic statements of *Theory X* organizations are described as:

    a. There is a critical, fault finding upper management;

    b. Hostile relationships exist within and between departments;

c. Work rates emphasize the attainment of individual achievements;
d. Emphasis is placed on whom to blame if things go wrong;
e. An independent inspection group must be used to catch defects;
f. Workers are not consulted for work related improvements;
g. Workers do not care about the company, products produced or quality.

The basic statements of *Theory Y* organizations are described as:

a. Upper management empowers workers to improve the work process;
b. Supportive, professional and friendly relationships are clearly visible;
c. Employees share incentives and individuals are encouraged to become leaders;
d. Self inspections are used to discover, rectify and prevent future defects;
e. Continuous improvement begins with soliciting worker feedback and suggestions;
f. Workers truly care about the company, the product and take pride in the quality level of their respective department.

So, which organization do *you* feel is better, more productive and can help you - *the manager* - achieve *your* goals?

The truth is many companies are run as *Theory Y* organizations but all the indications point to a *Theory X* company in reality. Some managers may see *Theory Y* as being too relaxed or not getting as much productivity from the workforce. Some managers lack the business and psychological skills needed to make a *Theory Y* organization successful.

The truth is, *Theory Y* yields more productivity, creativity in solving problems and a more stable workforce. When employees are empowered, they take a vested personal interest in the output of a process.

Passionate workers begin to develop a skill set that they can use in any department in which they work. Since the outcome is *"personal"*, they are invested in not only doing their job but also ensuring they give as much of themselves as the task requires.

Imagine if one hundred workers were empowered and shared your drive to see your goals achieved. Does this sound like something you could use in your business?

Management must begin to understand the need to truly know what employees want from their organization in order to align their goals with that of management. Without that vital link, managers may find themselves in a position of *"commander without an army"*.

A person must admit there is a problem before a solution can be found and implemented. Ignoring a workforce problem will not make it go away. It will, in fact, become much worse and may ultimately cause a total breakdown of the system. The manager may indeed face this consequence.

How do you motivate the workforce? Why should an hourly - paid employee care about helping you, the manager, to meet *your* objectives and goals? What would inspire a worker to ensure *your* goals are accepted as *his / her* goals?

In one word: *Motivation.* There are two types of motivation.

Intrinsic Motivation - the undertaking of an activity, as a hobby, without external incentive; also, personal satisfaction derived through self-initiated achievement[4]

Extrinsic Motivation - motivation that comes from outside an individual. The motivating factors are external, or outside, rewards such as money or grades. These rewards provide satisfaction and pleasure that the task itself may not provide.[5]

However, there is more to it than simply motivation. The most motivated employee will soon become frustrated and give up without the other very important - **mandatory** - tools to succeed. An employee must have *all* the necessary items.

The first item is *training*. Without training, you are paying the employee for *not doing their job*. Without training, employees will soon fail in their positions. Failure, one or many occurrences, will force the employee to withdraw and begin to feel isolated and embarrassed about the failures of which they unnecessarily accept ownership.

The employee must have the *tools* to do the job. This sounds very elementary but in the corporate world, the problem *does* exist. Basic items such as *reasonably* speedy internet access, ergonomic chairs and other general office supplies may be overlooked by upper management disconnected from the reality of their worker ranks.

Third, employees must *know* if they are meeting the expectations of management in the performance of their jobs. This is measured, in most companies, through a performance review methodology. This is a valuable process that is often an opportunity wasted when management and the worker do not see the benefit of this important formal evaluation.

Chapter 3

What are the attributes that all great leaders possess? Notice, I said *leaders* and not *managers.* There is a difference. Let us begin by defining *leadership* and *management.*

Leadership: *An essential part of a quality improvement effort. Organization leaders must establish a vision, communicate that vision to those in the organization and provide the tools and knowledge necessary to accomplish the vision.*

Management: *The act, manner, or practice of managing; handling, supervision, or controlling.*

It must be acknowledged that *"Every Leader is a Manager but not all Managers are Leaders".*

Leaders are those individuals who make an effort to better themselves in not only the areas of management but in the areas of human psychology and sociology. The leader knows how to manage but recognizes that managing is not enough to achieve his / her professional goals. The leader knows a title or position alone *does not* make a leader and wants to learn more about making their job easier; working smarter and not harder.

A leader is proactive and not reactive. Leaders make realistic goals and empower their workers to assist in the achievement of those goals. A leader knows that the worker will be the key to the leader's success or failure and responsibility to truly lead is not taken lightly.

Leaders encourage others with positive demeanor and a *"can do"* attitude. The leader is a *"people person"* because he / she truly understand the employee can be for or against them, depending on their perception.

Probably the most important attribute of leaders is their ability to align people with a vision; persuading and motivating them to accept the vision as *their very own.* This is not accomplished with huge pay raises or bonuses alone, but with intangible qualities called *trust* and *faith.* Many managers will never understand this concept and so they will always be a *manager* and never make the transition to *leader.*

Leaders recognize their limitations and work to bridge those gaps. There are those who claim to be leaders but their subordinates tell a different story. A person does not call himself a leader but a leadership role is a position which is assumed. There are traits a leader has which are learned and those which are natural attributes of the person.

The U. S. War College conducted a study of highly regarded major generals in 2005 and found their subordinates responded unanimously with the following feedback:

    1. A leader keeps cool under pressure;

    2. A leader explains missions, standards and

priorities;
3. *A leader sees the big picture and provides context and perspective;*
4. *A leader makes tough, sound and well thought-out decisions in a timely manner.*

This study also pointed to the fact that even when tactical and technical competence was excellent, the interpersonal skills were critical. It was noted that these skills were extremely difficult to teach and suggested they were easier if learned by example.

Leaders derive their roles from those who will follow them. It is this *"free will"* element that is the cornerstone of leadership. People may be ordered to follow by a manager but a leader simply points the direction and begins the journey. The leader is always in the front of the pack and clearly communicates the course changes to the group.

It is this *"information"* that keeps the workers aligned with the objectives. Without the communication, the followers will be unsure of their role and will begin to doubt the leader. Keeping employees informed frequently is time-consuming but the leader recognizes it is important and required.

Leaders are innovators. They develop their workers to ensure they are always comfortable with speaking up and adding their solicited thoughts, comments and suggestions. A leader knows he / she can not and *will not* succeed by themselves. It is as much about knowing their limitations as knowing strengths; both are important but limitations will lead to failures.

Leaders accept responsibility for the welfare and safety of their workers. It is this element that is seen by the worker and it is also this element that forms a bond between the two.

Again, the worker must have a personal investment in the outcome. This leads to a *"move mountains"* mentality that a manager will never enjoy. *Figure 4* illustrates the differences.

LEADERS	MANAGERS
Do the right things	Do things right
Focus on WHAT can be accomplished	Focus on HOW things should be done
Innovation	Conformity
Commitment	Control
Outcome oriented	Rules oriented
Transformational	Transactional
Energize the system	Ensure stability of the system
Vision, Inspiration, Courage	Procedure, Strategy, Objective
Create change – take followers from one place to another	Manage change – ensure the ability to handle it

Figure 4: Leaders vs. Managers

Of course, the *Leadership* role is always evolving and those who are flexible and embrace change will succeed in the business world of tomorrow. Obsolete views and assumptions will always be a challenge to the leader of the future.

One of the most difficult obstacles is communication. In a business world where email serves as the official communication plan, there will always be miscommunication and failures, internally and externally.

Leaders who rely upon such communication without a *"closed loop"* system will soon find the communication plan breaks down.

Communication remains one of the most important *"human"* elements and a failure at that level can have catastrophic effects on an organization.

Organizations with multi-level and / or multi-national challenges must view the communication plan as a *"high risk"* on their *Risk Management Profile*. A breakdown of the communications flow, or even worse, the absence of steady periodic communications, will inevitably lead to employee dissatisfaction, loss of efficiency and a systematic growth of mistrust from the frontline level to the executive offices.

This critical failure does not end there. Whether it is deliberate or not, this will begin to have an effect on the customer relationship. Without clear direction, the frontline employees will not be able to efficiently and effectively field the customer's questions and concerns, especially concerning their future relationship.

It should not be taken lightly when considering the avenue and methodology used to communicate with associates within a company. Management may try to justify the lack of communications by simply saying the information is *"strategic and / or proprietary in nature"* and divulgence is strictly on a *"need to know"* basis.

In the end, the *frontline* employees do not care about the reasons........*they care about the lack of information*........underline. Can you risk having a failure with *"lack of communications"* defined as the root cause? A *leader* understands this risk and chooses to lower the risk of occurrence through the employment of an effective, efficient and reliable communication plan.

A common view between leaders is the structure of the organization which they lead. Leaders understand the frontline employees are the ones who deliver the product of service to the customer. Without the employee, there will be no corner office, bonuses or organization to manage. It is this simple, *but most important* fact which dictates a different structure of organization.

The frontline employee is the one who will *"make or break"* the company / customer relationship. Leaders have structured the layers of their companies to facilitate supporting the frontline employee and this idea has resulted in a different hierarchy and business philosophy.

This view of the organization, in a visual sense, resembles an up side down triangle with the pinnacle being at the bottom instead of the top (*Figure 5*).

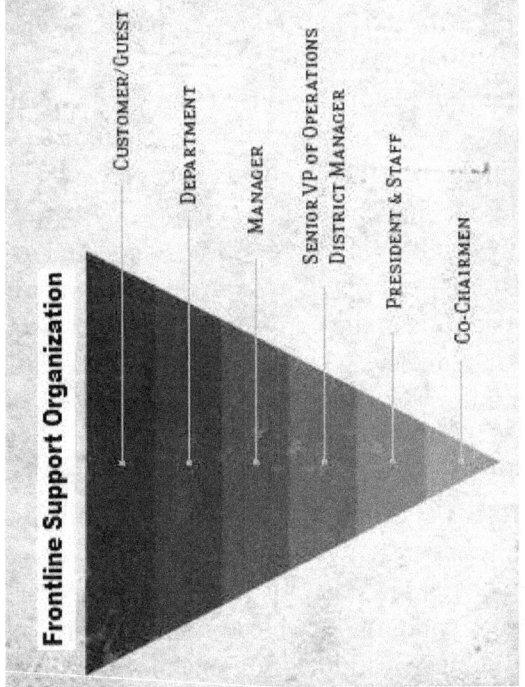

Figure 5: Frontline Support Organization

In this model, the support flows up the organization to the ones whose deliverables directly affect the customer. All decisions, directions and resources are strategically necessary to support the frontline employees.

Within this support system, the communications not only flow up from the leadership ranks but also flow down from the frontline as feedback and customer concerns. The more information collected from the source (*customer*) the better-prepared leadership will be with future decisions.

Knowing the customer will enable the company to exploit resources to the maximum and allow for the development of other products / services which the customer has indicated as *value added*. Anytime the company can provide those *value - added* choices to its customer's menu, the company is better positioned to beat the competition and increase their customer relationships and as a result, increase the bottom line.

Any organization must always strive to reject the *status quo* mindset. A company that begins to feel comfortable and invincible may soon find itself in bankruptcy court.

Even the deep pockets of a parent company may not help. Very soon, the troubled company may witness the financing and fiduciary assistance of the parent company "*drying up*".

Chapter 4

"Our plans miscarry because they have no aim. When a man does not know what harbor he is making for, no wind is the right wind."

Seneca

Planning is a major component of any organization and without objectives, all is left to chance. Amazingly, many corporations have adopted the *"let's see"* mentality which has resulted in their loss of the strategic edge and advantage. This may seem far-fetched but actions speak louder than words and the truth is revealed through those actions.

When asked what their objectives are, many organizations respond, *"To make money"*. This is a bit of a cliché, which does not represent a true objective.

There are many ways to make money for an organization; some *positive* (*"Be the best at fulfilling the customers needs"*) and *negative* (*"Cut salaries and jobs"*). Both will *"make money"* for the company but the *positive* objective is too vague and the *negative* objective could be devastating to the company.

Without clear objectives, different departments may waste resources or even compete with one another; which is counterproductive and a serious problem.

Objectives are important to not only lead the company to attainment but also to provide a way to get there, while measuring the progress along the way.

Objectives should be well thought out and a serious analysis of the company should be made during the process. For leadership to arbitrarily create an objective that *"sounds impressive"* or will *"make shareholders happy"* may very well be the *"nail in the coffin of their career"*.

Objectives must be created by leadership and cascaded throughout the organization. The objectives should be measurable (*achievement / timeliness*), attainable, easily understood, specific, realistic, in line with the company vision (*Figure 6*) and communicated to everyone in the company.

Figure 6: Organizational Objectives

While this seems like *Business 101* at the *Community College* level, *which sounds very elementary*, many companies have omitted these important planning issues. Objectives do not create themselves and from the CEO to the hourly paid employee, all should know, understand and work to attain the objectives.

Strategic planning represents the idea of forward - looking leadership and a workforce in step with the planned objectives. *Everyone* should know the objectives and the only way to know *exactly* where you are as a company, is to measure those objectives. It should be no secret as to the progress along the way. It must begin with the *Vision*.

Analyze your company at the *high level*. Answer the next few questions.

1. What is your company's *Vision*?
    - Is it easily understood?
    - Is it exciting to stakeholders / shareholders?
    - Is it clear and comprehensive in meaning?
    - Does it set the tone for employees?
    - Does it create unity of purpose?
    - Is it challenging and attainable?
2. What is your company's *Mission Statement*?
    - Is it your company's core purpose?
    - Is it inspiring?
    - Is it market-focused?
    - Does it reflect what *you* want the company remembered for?
    - Can it be measured?
3. What are your company's *Guiding Principles*?
    - Do the Guiding Principles provide the framework within which the company will pursue and achieve its *Mission*?
    - Are these values shared throughout the organization?

4. What are your company's *Broad Strategic Objectives*?
   - Are they general enough to prevent frequent revisions?
   - Are they easily understood?
   - Are they in the 5 - 10 year timeframe (*3 - 5 is actually better*)?
   - Do they support the Mission Statement?
   - Are they more specific than the *Mission Statement*?
5. What are your company's *Tactical Objectives*?
   - Will they achieve the *Broad Strategic Objectives*?
   - Can they be accomplished within a defined time period?
   - Are they quantifiable and measurable?

If you do not have all of these elements or you answered "*no*" to any of the questions, you may need to review, amend and implement as needed.

Strategic Planning[6] begins with collecting and arranging facts and data from the *Strategic Thinking*[7] process. The most commonly used tool by upper management to collect the information is the *SWOT Analysis (Organizational Strengths, Weaknesses, Opportunities & Threats)*. This tool helps to determine the company's position. The Strengths / Weaknesses are usually viewed as internal and the Opportunities and Threats are usually external.

For example, the limited cash flow of a company would be categorized as a *Weakness* while owning 85% of a market share would definitely be a *Strength*. Knowing trends in market growth will also be useful in preparing the SWOT.

A key concern when preparing a SWOT is a lack of objectivity. This may be as simple as making unfounded assumptions that are not supported with facts and data or downplaying a true *threat*. Intuition and hopes should not be used in preparing the SWOT. If you cannot measure it, you cannot accurately predict it. If the item may become a *Threat*, treat it as such to be safe.

Once the SWOT is complete, it is time to form the framework of the *Vision Statement*.

The Vision Statement

The *Vision Statement* is one of the most misunderstood of the Strategic Planning tools. The vision of the organization represents the *"dreams that can come true"*. The formation of the statement should include input from all senior leaders and it tells where the company is going in the future. It can be expressed as monetary, competitive or superlative and should be for at least a five-year timeframe.

An example of a Vision Statement: *"To become number one or number two in all markets we serve and revolutionize our company to have the strengths of a big company while combining the leanness and agility of a small company."* General Electric

The Mission Statement

If the Vision statement represents dreams that can come true, the *Mission Statement* goes one-step further to describe the organization, what it does and where it is going in the future.[8] The Mission defines the purpose of the company and is often combined with the *Guiding Principles* or *Values Statement.*

An example of a Mission Statement: *"Our purpose is to delight our customers by delivering a quality service, in a cost - efficient manner through the contributions of all associates."* Prudential Assurance Life

The Guiding Principles (Values)

These principles provide a methodology for the pursuit and attainment of the organization's *Mission.*

Usually constructed into five to seven core values, the Guiding Principles represent shared ideas and give shareholders and stakeholders alike, a representative summary or abstract of business, market, financial and in some cases, moral stance.

Example of Guiding Principles:

1. Our company places customer loyalty as our

highest priority, even above customer satisfaction;

2. *We will consistently exceed our customers expectations;*
3. *We will actively engage our employees in our business decisions;*
4. *We treat our stakeholders as partners;*
5. *Our company remains environmentally friendly and strives to operate with a passive impact on the environment;*
6. *Our company will treat our employees with dignity, respect and in a professional manner;*
7. *We will continuously improve our internal and external processes, procedures and relationships.*

Broad Strategic Objectives

A goal is usually long term (3 - 5 years) while an objective is short term. Objectives *"get the job done"*, so to speak.

The *Broad Strategic Objectives* are more specific than the Mission and they commonly fall into the realm of *"what"* instead of *"how"*. Considerable thought and planning should go into the creation of these objectives as to prevent frequent rewrites.

Strategic Objectives should be easy to understand and apply by all employees. They should represent a 3 - 5 year period, although 5 - 10 years is perfectly acceptable.

The objectives must also be aligned with the organization's *Guiding Principles* and they must support the *Mission Statement*.

As implied by the name, *Broad Strategic Objectives* are strategic by nature, which requires senior leadership to be intimately involved in the planning, construction, implementation and evaluation of the objectives. Once created, these objectives take on a life of their own. Without top - down support, these objectives will simply be *"warm and fuzzy"* words. No business has ever succeeded by sitting around the campfire singing *"Kum Ba Ya"* (*except maybe the YMCA*). Without actions, these words will not accomplish anything.

The use of *Scenario Planning* is also a good tool to assist with creating Strategic Objectives. When developing objectives, leadership may be prone to *Groupthink*[9], overconfidence or tunnel vision.

The *Brainstorming*[10] technique will help create scenarios that will logically consider *cause and effect* situations. Once the different scenarios are visualized, appropriate contingency plans can be developed. This works especially well when creating a *Risk Management Profile*[11].

Tactical Objectives

The last strategic planning level is represented by the *Tactical Objectives*. These objectives are well-defined methods for achieving the *Broad Strategic*

Objectives. It is usually at this point in which the focus begins to shift to allocation of resources, budgetary assignment and funding.

As the name implies, the tactics are described and the methodology defined. One of the most important attributes for the objectives is being *"measurable"*. It is very important to keep that in mind when creating the objectives. *"If it cannot be measured, it cannot be managed"*.

These objectives must be specific and detailed. There should be no misunderstandings within the organization and the draft of objectives should be cascaded through the organization for comments and/or feedback. If there are questions, or if the objectives are too vague, they should be amended before deployment.

This process may take several amendments and additional objectives may be created from the solicited feedback gathered.

Factors To Consider

There are certain factors to consider during strategic planning. There are many tangible and intangible elements that require thought and a thorough understanding before the planning phase is complete.

The entire planning phase may last six months or perhaps more depending upon the complexity of the organizational structure and market.

Professor Michael Porter of *Harvard Business School* developed a list of competitive forces, which should be analyzed to exploit the results of the SWOT analysis. Application of the list to the SWOT will help define the areas that need more attention and the ones that need adjustments.

The *Five Competitive Forces*[12] are:

1. *The threat of new market entrants* - This will have a direct effect on your business and planning the contingency is the best possible mitigation method. The six possible barriers should be considered:

   - *Economies of Scale*: A new entrant into the market must be prepared to compete on a large scale which requires exceptional operating techniques;
   - *Product Differentiation*: If brand loyalty is an obstacle, this may force new entrants to invest substantially to counter brand loyalty effects;
   - *Capital Investments*: Large investments in inventory, marketing, R & D, etc. may be required;
   - *Learning Curve*: This cost advantage may occur from being further down the learning curve and having assets such as experience and intellectual property rights;

- *Distribution Channels*: Existing distributors may be closed or open to new entrants;
- *Government Policies*: Regulated industries may enjoy protection from new entrants but changes in regulation may create new entrants who can fill a void or vacuum.

2. *Power of Suppliers* - A supplier will have a better bargaining position if:
   - The industry is dominated by fewer companies;
   - The supplier's product is unique;
   - The industry is not of major importance to the supplier;
   - There are no substitutes for the product.

3. *Power of Customers* - Customers and suppliers are considered to be on opposing sides in an industry. The customer becomes powerful if:
   - Economies of Scale are large and purchases are large;
   - The product is a small part of the buyer's total cost;
   - The buyer is in a "*low cost*" industry and must buy accordingly.

4. *Substitute products* - A cap on potential profits may result and cause price reductions throughout the industry.

5. *Industry rival* - Significant competition may create price wars and collapse the market

profits. Industry rivals have the following characteristics:

- Numerous market competitors;
- Slow industry growth;
- Product is a commodity;
- Excessive capacity;
- The *exit barriers* (cost of leaving the industry) are too high;
- There is intense rivalry.

A company can use this tool to better position itself in a market, protected by the capabilities and advantages, and it can reduce the threats while taking advantage of future business opportunities.

One other consideration is the stakeholder. The objectives and goals must be aligned with the needs of the stakeholder. A stakeholder may be stockholders, leadership, employees, suppliers and customers. ISO also recognizes the community and society as a stakeholder.

It is important to consider these groups collectively and individually to ascertain the needs and ensure the decisions address their needs.

Quality Management Integration

Chapter 5

Audits are an important part of the overall *Quality Management System.* The audit will not only verify compliance and/or conformance, but it will also identify areas of improvement and ways to streamline the business. The audit represents the best way for management to get a good idea of how the audited department is truly functioning.

In the past, audits have been viewed but the Auditee as a punitive measure that uncovers mistakes and could perhaps result in disciplinary actions. The auditor was someone to be feared, loathed and avoided if at all possible. The view in the modern business world contradicts the negative impressions and misconceptions of days past. Management today realizes the advantages of audits and uses the audit as a management tool for everything from process improvement to verification that perceived additional resources are in fact, needed.

As Quality has evolved, the audit has taken on a different appearance in its performance. The *"police – like"* audit interviews of the past have made room for a more guided discussion and conversation, where the information is exchanged and verified by the auditor. The audit now focuses on conformance as opposed to non-conformance.

The verification of audited material remains, for the auditor, the charge of his / her due diligence. There are, however subtle differences in what the auditor is really doing to accomplish the goal of the audit performance. The auditor focuses on identifying opportunities for improvement and ways to bring savings to the business.

Another popular and cost / time effective method is the use of inspections. The inspection is a limited scope audit that examines only a couple areas of a process or procedure to verify the final product or service meets the specification of the internal and external customer.

The inspection takes less time and allows the auditor to verify compliance to the standard while allowing the exploration of ideas to reduce waste and increase the speed of the process without compromising the quality of the process output.

The importance of feedback to management from audits cannot be overstated. The data gathered and delivered to management is invaluable in setting goals and objectives. To ignore the information is to abandon management through facts and data to rely solely on *luck*, which runs out soon or later for everyone.

The collection of qualitative and quantitative metrics is important in business analysis and trend analysis. Quantifiable *trigger points*, strategically place in the metrics, will give management the opportunity to *see into the future* and prepare for approaching risks before they come to fruition. *Corrective*[13] and / or *Preventative Actions*[14] result from audit performance. These tools are designed to improve the process or system using *Root Cause*[15] evaluation, implementation of a *"fix"* and prevention of future defects of the similar nature.

Corrective action plans are developed by the *Auditee*[16] or responsible manager. It should clearly identify the necessary activities that should be implemented.

A documented process should exist for dealing with *Findings*[17] that result from an audit. The process should include assignment of responsibilities for short term, immediate action to contain a product, process or service *nonconformity*[18].

Some corporations have committees (*Corrective Actions Boards*) that coordinate the corrective / preventative action activities. The quality department should always be involved in reviewing the actions through *follow-up*[19] to ensure the measures are affective. Quality must remain objective during the evaluation.

Most chronic problems can be solved by simply trouble shooting to find the root cause. A typical (and simple) procedure may follow the following sequence:

1. Assign responsibility for actions;
2. Evaluate the importance;
3. Investigate possible root causes;
4. Analyze the problem;
5. Create corrective / preventative actions;
6. Follow-up to ensure actions are effective.

Corrective actions usually fall into one of the following categories:

- Immediate Actions: Actions taken to stop the problem immediately. *"Would you want defects to continue to be sent to your customers?"*

- Temporary Actions: Actions taken to stop the problem in the near term. *"Do we shut down the assembly line and inspect 100% of the product until the run is complete?"*

- Permanent Actions: Actions taken to stop the problem forever. *"Shut down the line, overhaul the machinery and start the line again. We will inspect 100% of the product until we meet Six Sigma[20] (99.99%) standards."*

When a finding is discovered through auditing, the most critical action is to evaluate the risk resulting from the output of the process. You would not want to discover a *widget* is out of specification while one hundred *widgets* are loaded on a truck and sent to your customers. A quick *risk analysis* must be completed to see what the extend of damage (*defects*) may be present.

A *process analysis* should also be conducted to ascertain if any processes that touch the defective process have been or will be affected. The process analysis should be followed with a root cause analysis to determine what caused the defect. If the defect occurs in a process that produced 500 widgets per hour, the scrap costs could be astronomical, not to mention if a defective item is allowed to be shipped to your customer.

Auditor follow-up can be one of the weakest links in the audit process. The auditor must evaluate the implemented changes and objectively consider if the actions are effective. If the actions have a *"reasonable opportunity for success"*, the auditor should accept the actions and evaluate during future audits and inspections.

After follow-up and verification, the item may be closed. The process should address the procedure and documentation required for closure.

It is also important to remember auditors cannot "fix" their findings. If they are involved in the correction, they should not conduct the follow - up to close the finding. Auditors cannot audit their own work.

Quality Management Integration

Chapter 6
"If you can't measure it, you can't manage it"

Fred Smith

The point of having a QMS^{21} is to assist in producing the metrics needed to manage the company. If these metrics are not produced, or even worse - not used, management will soon find themselves *"fire fighting"* instead of managing the business.

$Metrics^{22}$ tell a story and will even tell the future, if interpreted by an objective, prudent manager. Would you run a company without working closely with your finance department? How would you know where you are and where you are going financially? How would you know if you can afford to expand the business and what indicators would you have of an impending financial failure of the company?

The operations area of a company is no different. You must keep your finger on the *proverbial* pulse in order to know the health of the operation. Without this important tool, you may find your company employment longevity is defined in *days - not years*.

However, be aware: the information and how it is interpreted, is more important and can actually have a negative effect if the *take away interpretation* is wrong. Remember I said *objective, prudent manager*. Objectivity is paramount and prudence is closely following.

Quality metrics can be made easier to understand by using tools that enable the manager to make an educated and informed decision on how to proceed and direct the business. The easier the information is to understand, the easier it will be to correlate into the decision - making process.

Let us take for example, the *Pareto Chart*[23]. This chart is a specialized column graph used to prioritize problems so *major* problems can be easily identified. The Pareto gives a clear illustration of the serious issues. It is used to separate the *"vital few from the trivial many"*[24]. Briefly stated, the Pareto suggests a few categories (*approximately 20%*) represent the most opportunities for improvement (*approximately 80%*).

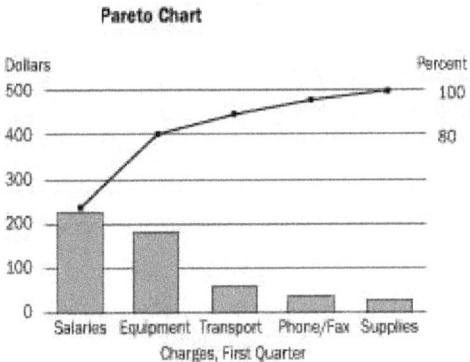

Figure 7: Pareto Chart

Pareto charts are helpful when used to:

- Analyze a problem from a new perspective;
- Focus attention on problems in priority order;
- Compare data changes during different time periods and;
- Provide a basis for the construction of a cumulative line.

The Pareto uses *"first things first"* linear thought to interpret the information presented in the chart. The viewer is immediately attracted to the priority order of importance. In this fashion, upper management can see the most pressing issues and when the attention needs to be *immediately* focused.

The *Histogram*[25] is a graph characterized by the number of data points that fall within a given bar or interval, which is commonly referred to as *"frequency"*. The Histogram usually requires a minimum of fifty to one hundred data points to capture the measurement of the process in question.

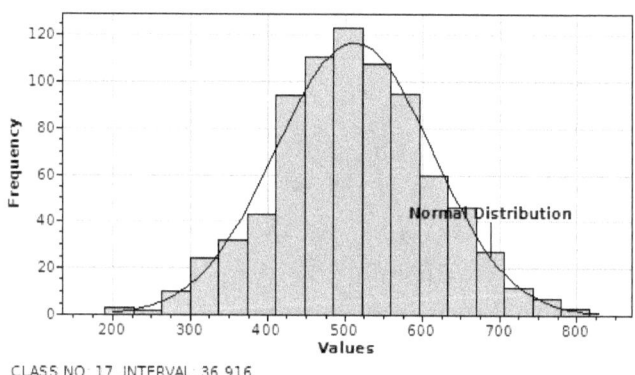

CLASS NO: 17, INTERVAL: 36.916
Samples: 800, Sum: 4.0884E5, Min: 189.33, Max: 816.91
Mean: 511.05, StdErr: 3.5618, Std[N]: 100.68, Std[N-1]: 100.74
Skew Gs: 0.003, Kurtosis Gk: 3.055

Figure 8: Histogram

An *Ogive*[26] displays *running totals* data, also known as *accumulated frequency*, which can be illustrated by a *Stem and Leaf*[27] distribution.

Table 1

Stem-and-Leaf Plot Showing Distribution of Adult
Heights (N=200)

Inches   .1 inches

55 3
56 9
57 7
58 57
59 25
60 228899
61 1225667
62 033556899
63 011223558888999
64 012333344555688
65 1123345555566788889
66 0112234455668889999
67 000122234555566666667899
68 11222333344445666799999
69 112335555589
70 1244566779
71 234455789
72 11223344666778
73 06678
74 8
75 18

Figure 9: Stem and Leaf Plot

Once collected, the data points can be plotted on the
Histogram and a correlation between the points

established. If these data points exhibit a unimodal or *bell-shaped curve*, the process is stable and a stable process is predictable.

Histograms are utilized as *Control Charts*[28] and parameters can be established to ensure the data points remain within the defined criteria. The data set may be compiled into a bar representation and a cumulative or trend line added to the chart.

Control Limits[29] are added to the chart to quickly identify any data points that fall outside the established parameters. When a data point exceeds the *UCL / LCL*[30], it has demonstrated a characteristic of a unstable process. A *Truncated Histogram*[31] reduces the limits of the data sets to a more focused area of possibilities.

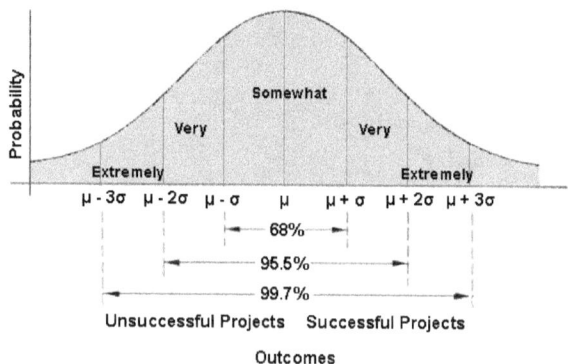

Figure 10: Bell Curve

There are many distributions that do not follow the normal bell curve. Some examples are the bimodal, exponential, lognormal, rectangular U-Shape and Poisson[32] distribution.

Scatter Diagrams[33] (*correlation charts*) display relationships between data points and may represent *Low-positive*, *High-negative* or *Non-linear* relationship characteristic displays. These diagrams are created along X and Y-axis orientation. A line can be charted along the closest data points to form the relationship.

Scatter Diagram Interpretation

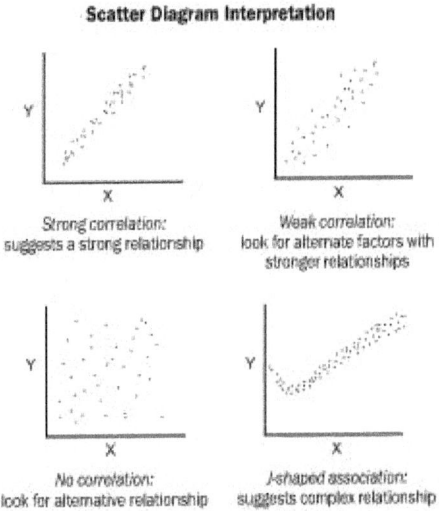

Strong correlation: suggests a strong relationship

Weak correlation: look for alternate factors with stronger relationships

No correlation: look for alternative relationship

J-shaped association: suggests complex relationship

Figure 11: Scatter Diagrams

One of the most useful tools to determine causes for undesired events is the *Cause and Effect Diagram*[34]. This diagram:

- Breaks problems down to smaller, manageable elements;
- Displays possible causes in a graphical manner;
- Also known as a Fishbone, 4M or Ishikawa diagram;
- Illustrates how various causes interact and;
- Used with the *Brainstorming* technique.

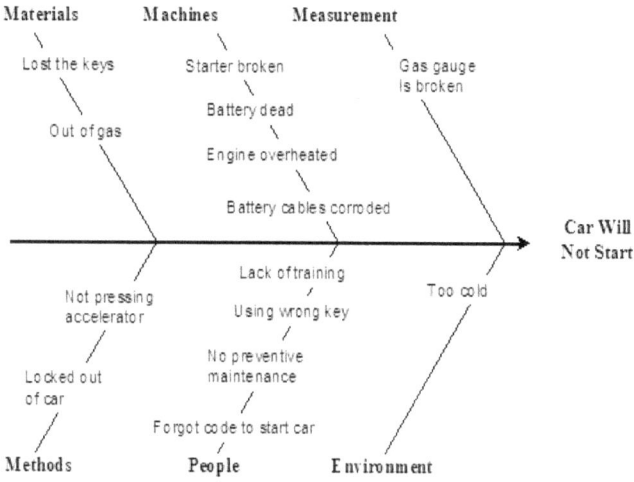

Figure 12: Cause and Effect Diagram

Flow Charts[35] are the most common process-mapping tool in modern business applications. *Action-to-Action* charts are flow charts that depict the sequential flow of activities within a process. It is a good tool for determining how the process works and also identifying areas for process improvement.

It is of utmost importance that processes be documented and analyzed by management *before* deployment and personnel's training occurs. This will negate the necessity of retraining due to a total amendment of the process in question.

The process analysis should also consider how the process integrates into the overall system. Every process will have inputs and outputs and usually function in concert with other processes. If one process is flawed, the entire system may be at risk.

Quality Management Integration

Chapter 7

Since the first applications of quality principals, the question of *"How much does quality cost and who pays for it?"* has ricocheted through the corporate management ranks. The opinions vary greatly.

Some operations managers view quality as a shared service, which must be budgeted across departments, which shares the expense of quality. Other managers are quick to point out their departments are audited only a fraction of the audits other departments undergo and reason *their* expenses should be less than those departments requiring quarterly, monthly or weekly audits or inspections.

The *Cost of Quality*[36] *(COQ)* is often measured by the *Cost of Poor Quality*. Cost of Poor Quality *(COPQ)* is the cost associated with providing poor quality products or services.

There are four categories:

- *Internal failure costs* (associated with defects found before the customer receives the product or service);
- *External failure costs* (associated with defects found after the customer receives the product or service);
- *Appraisal costs* (incurred to determine the degree of conformance to quality requirements) and;

- *Prevention costs* (incurred to keep failure and appraisal costs to a minimum).

While these measures can be easily quantified, other costs of poor quality are not so easily defined and measured and can, in some cases, be easily hidden. Examples are:

- Unreported scrap (*muda*[37]);
- Waste added intentionally to budgets (over - inflation of budgets) and;
- Dissatisfied customers.

The most important and *critical* hidden cost is customer dissatisfaction, which can only be measured when the customer *"votes with their feet"* and chooses your competitor.

The loss of revenue can only be traced to poor quality if other factors and information are considered and a root cause analysis performed. For these reasons, management may never consider the loss of the customer to be a true cost of poor quality.........*yet it is the root cause.*

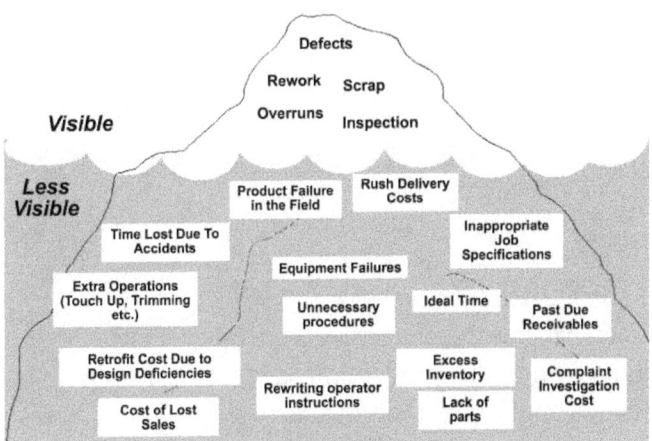

Figure 13: Visible vs. Less Visible Costs
(Cost of Poor Quality)

As illustrated in *Figure 13*, the *COPQ* can result in expenses that may or may not be easily defined. If a root cause analysis is not performed, some costs may be *written off* without fully understanding the true nature of the cost.

Some authorities believe that each *$1.00* spent on prevention costs will save *$3.00* in appraisal costs and an overwhelming *$7.00* in failure costs.

Some examples of prevention costs are:
- Capability studies;
- Forecasting;
- Job descriptions;
- Maintenance;
- Personnel reviews;
- Documented procedures / processes;
- A frequent review of company procedures / processes;
- Training / education;
- Vendor Surveys and;
- Customer surveys.

This list is not exhaustive and other costs may be placed into the prevention category depending on the accounting method in use.

Examples of appraisal costs:
- Audits;
- Equipment calibration;
- Inspections;
- Testing and;
- Shipping / receiving validation and inspections.

While *internal* failure costs are associated with the work itself, the *external* failure costs should be considered the most important; even if those costs are less than internal failure costs. The reason: *Your customer has experienced the failure!*

Again, it may be difficult to quantify external failure costs unless your drill down to the root cause. It is however, the most critical and can easily put your company out of business, regardless of which column your company puts the costs. It does not matter where the cost falls on the balance sheet if all your customers leave your company.

Activity Based Costing (ABC)[38] is a popular method of quantifying *COQ* and *COPQ*. The methodology can be easily implemented into an organizational structure and reported to upper management. There is one caveat: the costs must be related to a *specific* product, service or process. For that reason, many companies have a difficult time implementing *ABC*.

Companies that have implemented *ABC* have often discontinued some products or services simply because those products or services lost money, were unprofitable and were *"eating away"* at the profits of other, stronger products / services.

The *ABC* method contrasts traditional costing methods whereby the costs are based on an arbitrary percentage of direct labor costs. With direct labor costs becoming less of a factor in some products and services, some organizations no longer find the traditional method viable.

It should be mentioned that when implemented, *ABC* can make it easier for a manager to budget and receive quality cost dollars since they are tied directly to the proportion of output; which contributes to the company's *bottom line*.

Every system has pros and cons and *COQ* is no different. As mentioned before, information is only as good as the person interpreting the information.

There are some advantages and disadvantages associated with quality costing systems. Let us look at some examples:

Advantages:
- Single overview of quality;
- Aligns quality to company goals and objectives;
- Problem prioritization, especially when presented to management in a Pareto chart format;
- Accurate distribution of quality costs to ensure maximum profit;
- Effective use of company resources;
- Reinforces *DIRTFT*[39] principles and;
- Assists with establishing new product / services processes.

Examples of disadvantages:
- Does not *solve* the quality problems;
- Does not suggests specific action and corrections;
- Susceptible to short term mismanagement;
- Important costs may be omitted;

- Inappropriate costs may be included and;
- Susceptible to management interpretation errors.

While the *COQ* and *COPQ* will continue to be argued in boardrooms around the world, the matter must be considered and accounted for in the business.

Quality Management Integration

Chapter 8

"A ship in harbor is safe - but that is not what ships are for."

John A. Shedd, *Salt from My Attic*

Risk Management: *Using managerial resources to integrate risk identification, risk assessment, risk prioritization, development of risk handling strategies and mitigation of risk to acceptable levels.*

Risk is a part of business and everyday life. Taking a chance on a new project or product may pay big dividends for your business, but it comes with a price….that price is *risk*.

The uncertain circumstances relating to the unknown are often overwhelming for a company to *"wrap their arms around"* and this will lead to missed revenue opportunities due to the proverbial *"cold feet"* syndrome.

Risk does not need to be anything other than what it is: *uncertainty*. While the future is full of uncertainty, there are some methods that can help illuminate some of the unknown issues, plan for the contingency and address the risks that may occur. Risks can be negative *or* positive.

As with any initiative, planning is the most important aspect and many problems can be averted by proper, thorough planning. Risk does not need to be *"risky"*.

Some organizations and stakeholders are willing to accept varying degrees of risk, which is called *risk tolerance*. Some risks may be accepted if they are within the tolerances and are within the balance of rewards that may be gained by taking the risk.

Certain attitudes toward risk can be adopted by organizations and stakeholders that can influence the way responses to risk are handled. Such attitudes are driven by perception, experiences and personal biases. A consistent approach to risk management should be developed and adopted by a company to ensure uniformity across the entire departmental organizational structure.

For risk management to be successful, the company must be committed to a *proactive* approach, which identifies risks through all levels of the organization. Risks always exist and the proactive focus increases the possibility that a risk can be minimized or illuminated. The risk does not cease to exist simply because the company refuses to acknowledge and / or deal with the risk.

From a process aspect, risk management should be approached with the same careful and explicit planning, which will greatly enhance the probability of success. The type, degree and visibility of risks are commensurate with both the risks and the importance to the organization. Planning also permits the opportunity to secure proper resources and time for the risk activities and evaluation of risks.

There are certain risk attributes that define how an organization will address risk management:

- *Risk Averse* – Uncomfortable with risk;
- *Risk Neutral* – Embraces risk;
- *Risk Seeker* – Views risk as a challenge and;
- *Risk Tolerant* – Chooses to ignore risk and acts surprised when risk occurs.

From the attributes, companies will define their position when it comes to risk. Some organizations view risk with a *"zero tolerance"* approach. This organization is usually a large established company that does not want to take chances, regardless of the possible rewards.

The new upstart may base their business model on taking calculated risks in order to break into a new market or compete with a company that has a stronghold on a market segment.

An investment company that aggressively buys and sells volatile stocks may mandate the risk seeker approach as a way to outperform competitors. The company that chooses to *"hope for the best"* will resemble the risk tolerant attribute. Would you want your Financial Planner to *"hope for the best"* with your investments?

The best way to begin adopting risk management is support from upper management and a strong *Risk Management Plan (RMP)*. While the *RMP* is widely used in *Project Management*, it also provides a methodology for companies beginning to use the risk management philosophy. The *RMP* contains elements that provide a basis for the company to use.

The *RMP* contains:

- *Methodology*: This element defines the tools and data sources that will be used to perform risk management.
- *Roles / Responsibilities*: The individuals that will lead and support risk management and their responsibilities.
- *Risk Budget*: Assignment of resource funding to track the costs and payoffs of risk management.
- *Risk Review Schedule*: When, who, where and how often are risks evaluated and the protocols that are used.
- *Risk Categorization*: A documented method for quantifying and qualifying risks. This may take the form of a *Risk Breakdown Structure (RBS)*. The RBS is a hierarchical organized depiction that lists the categories and sub - categories within which risks may arise.

- A *Risk Rating* is often assigned by using the following formula:
  Probability * *Impact* = *Risk Rating*
- *Definitions*: This may contain all the definitions that are used in the organization's risk management processes.
- *Probability / Impact Matrix*: The risks must be prioritized according to their implications for affecting the company's objectives. This may or may not be part of the *RBS*. The risk attributes should be reviewed in detail with upper management.
- *Reporting Methodology*: Reports will be produced as a result of risk management processes. Documents, communications, analysis and the format of each will be defined in this section.
- *Recording / Tracking*: Documentation should be created and retained for the risk management processes including lessons learned, future needs foreseen and how the risk processes will be audited and evaluated. This will also include a *Risk Register* or *Risk Profile*. The risks will be updated as necessary according to their potential impact in the future.

Risk identification is an iterative process due to the fact risks are constantly evolving with the organization. What was not viewed as a risk six months ago may very well be the risk that will force your company into bankruptcy or out of business. Since business is dynamic, risk management must, *by nature*, also be dynamic. The identification of risks will be an evolution and a continuous cycle of objective evaluations and resulting actions.

Once identified, the risk must receive a ranking (*weighting*) evaluation to assess the impact and the aggressiveness of the mitigation measures. The first step should be a qualitative analysis. This process prioritizes the risks for further analysis or actions by assessing probability and the resulting impact.

The risks are ranked according to their severity and the resulting issues caused. An alphabetic paradigm is most commonly used and the analysis is usually a rapid and cost-effective means of establishing the priorities and prepares the foundation to quantitatively assess the risk.

Quantifying risks is the process of numerically analyzing the effects the risk may have in the future. This method also allows management to evaluate the risk with objective data instead of subjective opinions. When used together, qualitative and quantitative analysis ensures the risk is appropriately weighted and dealt with accordingly. Some tools for qualifying risks include:

- *Probability and Impact Assessment*: This

tool investigates the likelihood that each specific risk will occur and the potential effect on company objectives. A *Probability and Risk Impact Matrix* is often used to record the assessment. A *Risk Rating* helps guide risk responses.

- *Risk Data Quality Assessment*: Evaluation of the degree to which the data about risks are *useful* for the management of the risks. Does this sound *confusing*? In reality, this assessment simply evaluates the *reliability* of the data information gathered before making a decision based upon the information.
- *Risk Categorization / Urgency*: As discussed, this makes it easy for management to assess the risks at a glance.
- *Expert Judgment*: This technique is helpful in looking at the past and comparing it to the future. An educated guess can be made and applied to risks of similar nature and attributes.

Some tools for quantifying risks include:

- *Probability Distributions*: This tool is frequently used in modeling and simulation (*which can be used alone or with probability distributions*) representing the uncertainty in values such as duration and costs of activities.

- *Sensitivity Analysis*: This technique uses the *Tornado Diagram (Figure 14)* to compare relative importance and impact of variables that have a high degree of uncertainty to those that are more stable.

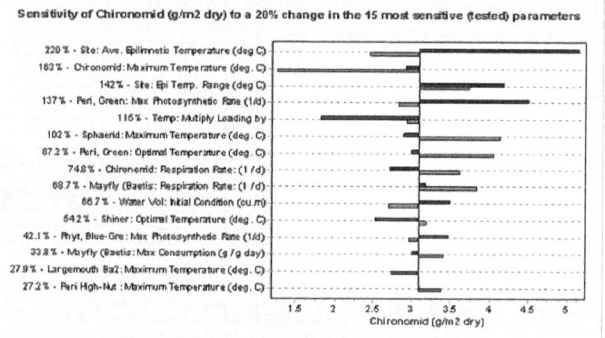

Sensitivity of Chironomid (g/m2 dry) to a 20% change in the 15 most sensitive (tested) parameters

Figure 14: Tornado Diagram

- *EMV Analysis*: Expected Monetary Value Analysis calculates the average outcome when the future includes scenarios that may or may not happen and requires a risk neutral assumption that is neither risk averse nor risk seeking in nature.

- *Expert Judgment*: This technique is helpful in looking at the past and comparing it to the future. An educated guess can be made and applied to risks of similar nature and attributes.

All of these planning techniques are of no benefit if

strategies are not developed to address the risks.

First, let us look at the strategies for negative risks.

- *Avoid*: Changing the company's plans or objectives to eliminate the threat of the risk. This may be in the form of upper managements' decision to discontinue a product of service to avoid the competition's price war.
- *Transfer*: Shifting some or all negative impact and ownership of a threat to a third party. Purchasing insurance for a supplier's shipment would be an example.
- *Mitigate*: Reducing the probability of occurrence or impact of the occurrence to an acceptable threshold. Designing redundancy into a process or system is an example of this strategy.
- *Accept*: This strategy can be either passive or active. Passive acceptance requires no further actions but it leaves the company to deal with the risk as it occurs. The Active acceptance involves planning such as a contingency of time, money or other resources as appropriate to handle the risk when it occurs.

Positive risks have an effect on the company viewed as *"too much of a good thing"*. An example of a positive risk could be the overwhelming success of a product. Due to the huge demand, deliveries, logistics, supplier issues and inability to meet market demands could result. It is great to have one million orders but what if you are not *equipped to fulfill those orders?*

Now, let us look at the strategies for positive risks.

- *Sharing*: Sharing a risk with a third party may be more beneficial than dealing with the risk alone. An example is a joint venture or partnership, where the expertise of one party makes up for the lack of the other.
- *Exploitation*: *"When given lemons, make lemonade".* This strategy capitalizes on the positive risk and makes the most of the situation.
- *Enhance*: Enhancement allows the positive risk's possibility of occurrence to be increased. This method requires opportunities to be realized by ensuring the risk happens.
- *Acceptance*: Being able and willing to take advantage of the risk if it happens, but not actively pursuing it.

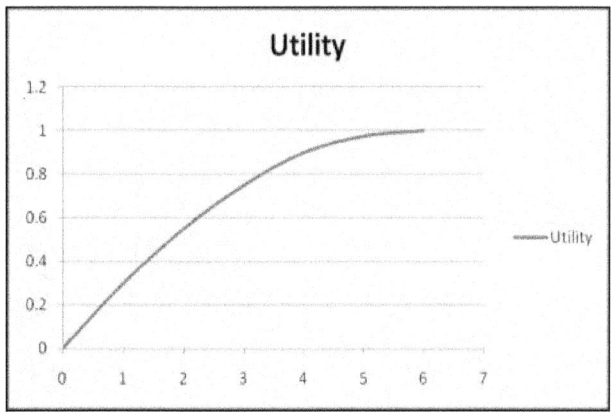

Figure 15: Risk - Averse Utility Curve

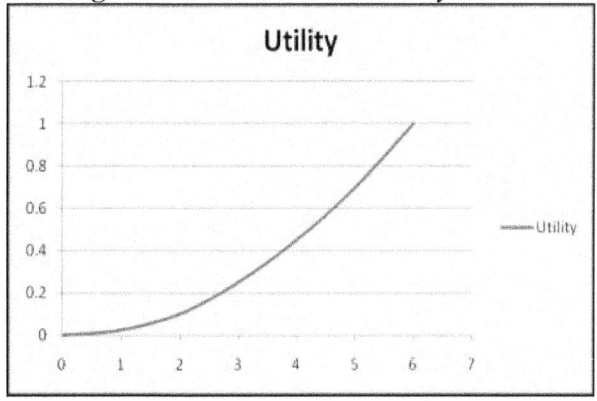

Figure 16: Risk - Seeking Utility Curve

Quality Management Integration

Chapter 9

In theory, *supplier*[40] (*also known as vendor*) quality management is fairly straightforward: *Deliver what is needed, when it is needed and to the specifications needed.* This sounds simple enough but in practical application, it is much more difficult.

An organization must recognize the limitations, strengths, weaknesses and other variables that will set the stage for suppliers to meet their expectations consistently. A rudimentary supplier quality program may suffice if the supplier is only one of a hundred *"off the shelf"* suppliers. However, if your business is dependent upon the success of your supplier, the stakes of the game are considerable higher. Are you willing to *"bet your company"* that the supplier will always meet your requirements?

First, a company must understand that supplier management is a *"relationship"* and like any relationship, good communication will form the cornerstone of the metaphoric foundation. Communication must be *constant* and the flow of information must be easily *understood* by all parties involved.

This may seem like overkill in the corporate world today but without it, time and money will be wasted on rework and failures; *internal and external.* The better the communication plan is designed, the better the results will be and the more saving can be realized.

Let's look at a simple case study as a means to understand the complexity of the principles involved in supplier quality management.

> *Company A* has just signed a contract with *Company B*. *Company A* will provide five thousand widgets per month to *Company B*. *Company B* will use the widgets in its final assembly of a hydraulic motor. *Company B* then sells these hydraulic motors to its customers who, in turn, use them on their offshore oil drilling platforms around the world.
>
> *Company A* has two iron ore suppliers (*Company C* and *Company D*) who supply the iron ore needed to manufacture the widgets. The iron ore grade, composition and quality have been established through existing written specification documents signed by the suppliers and *Company A*.
>
> *Company A* will need four shipments per week, two from both suppliers, in order to meet the manufacturing demand. Both suppliers have provided a consistent supply of iron ore in the past but have not had such stringent time requirements in place during the current agreement.

Company C and *Company D* usually purchases their raw iron ore material from the same supplier, at two separate geographical locations. The material is then delivered via railway to their respective plants, where the refinement process is conducted. Once complete, the iron ore is transported via railway to the manufacturing plant of *Company A*.

With the existing supplier agreement in place, *Company A* receives its first shipments from *Company C* on time and to its specification. This allows *Company A* to manufacture six hundred of the required five thousand widgets on time and on budget.

Company A receives the shipment from *Company D* three days later. Part of the shipment has been exposed to rainwater that leaked into the railway car, rendering the load *"unacceptable"* during routine sample inspections conducted by *Company A*. The load is therefore rejected. The remaining accepted delivery will only produce four hundred widgets. It is projected that *Company A* will only produce half of the required widgets during the first month of the contract.

The situation worsens when the employees of the supplier threaten to go on strike due to a breakdown in collective bargaining contract negotiations. This would affect *Company C* and *Company D* and would ultimately interrupt the supply to *Company A,* resulting in a material breach of contract with *Company B.*

Supplier management can be the most important part of the business and can result in the success or failure of the business. As illustrated in the above case study, *Company A* had a supplier program in place but failed to amend the agreement when the requirements changed. The attributes of grade, composition and quality remained important requirements but the *"time"* requirement became equally important due to the nature of the contract.

The failure of *Company A* to reevaluate the relationship with its suppliers quickly resulted in loss of revenue, legal issues and a lost customer in the process; the latter of which cannot be measured in a dollar amount.

There are several supplier quality models used in business today which are designed to meet the particular needs of the user company. There are many good books and internet articles on the subject which can help the decision maker access their operation and choose the appropriate system to implement.

It is not the intent of this book to delve into the various attributes of the numerous systems however, we will look at two supplier methods.

One system will provide benefits to a company *beginning* a supplier quality assurance system[41] at one sustainable level (*Supply Chain Management*) and the other *evolves* from the stable supplier system and elevates supplier management to a new level of confidence and monetary savings (*Just In Time*). Combined, the two systems can increase confidence, reliability, dependability and decrease costs.

Supply chain management (*SCM*) is the management of a network of interconnected businesses involved in the ultimate provision of product and service packages required by end customers (Harland, 1996).

The APICS[42] Dictionary defines *SCM* as the *"design, planning, execution, control, and monitoring of supply chain activities with the objective of creating net value, building a competitive infrastructure, leveraging worldwide logistics, synchronizing supply with demand, and measuring performance globally."*

A supply chain[43], as opposed to *supply chain management*, is a set of organizations directly linked by one or more of the upstream and downstream flows of products, services, finances, and information from a source to a customer. Managing a supply chain is *"supply chain management"*.[44]

Supply chains enable companies to improve their overall competencies in the same way that outsourced manufacturing and distribution has done; it allows them to focus on their core competencies and assemble networks of specific, *best-in-class*[45] partners to contribute to the overall value chain itself, thereby increasing overall performance and efficiency.

The ability to quickly obtain and deploy this domain - specific supply chain expertise without developing and maintaining an entirely unique and complex competency in house is the leading reason why supply chain specialization is gaining popularity.

The American Customer Satisfaction Index (*ACSI*)[46] continues to highlight these profitable relationships and credit them for high customer satisfaction ratings time and again.

Another advantage of SCM is accountability. The responsibility lies with the supplier and financial incentives (*for above and beyond*) and penalties (*for poor contract performance*) are often tied to the supplier agreement. This transfer of the risk to the supplier(s) is very attractive to many companies who do not wish to create an in-house support infrastructure, which detracts significantly from their bottom line.

Successful SCM requires a change from managing individual functions to integrating activities into key supply chain processes. Supply chain business process integration and *Process Performance Management*[47] involves collaborative work between buyers and suppliers, joint product development, common systems and shared information.

Some of the key processes identified by Douglas Lambert[48] are:

- Customer relationship management
- Customer service management
- Demand management
- Order fulfillment
- Manufacturing flow management
- Supplier relationship management
- Product development and commercialization
- Returns management

By formalizing these processes, it is possible to integrate the elements needed for successful supplier management and ensure the success of the relationship. The idea is to form the working relationship, which will strive for continuous improvement while streamlining the operations of both supplier and customer.

This mutual benefit is very attractive in the volatile business environment today and many companies have realized SCM will provide a modicum of stability and expectation in their respective companies, even during the most difficult of market situations.

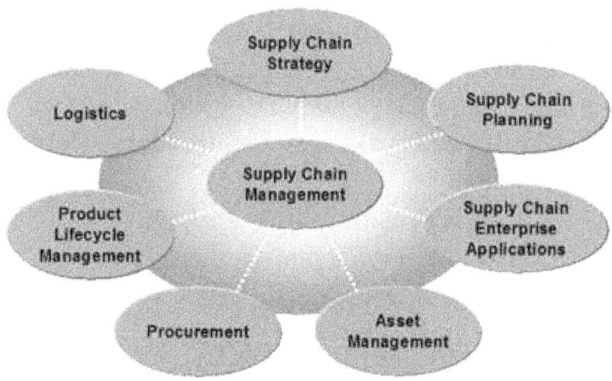

Figure 17: Supply Chain Management

Just In Time (*JIT*) is also known as a *"Pull"* (*demand*) driven inventory system in which materials, parts, sub-assemblies, and support items are delivered just when needed and neither sooner nor later. This delivery time is sometimes set at steady time intervals called *Takt Time*[49]. Its objective is to eliminate product inventories (*muda*)[50] from the supply chain.

As much a managerial philosophy as an inventory system, JIT encompasses all activities required to make a final product from design engineering onwards to the last manufacturing operation.

JIT systems are fundamental to time based competition and rely on waste reduction, process simplification, setup time and batch size reduction, parallel (*instead of sequential*) processing, and shop floor layout redesign.

Under JIT management, shipments are made within rigidly enforced '*time windows*' and all items must be within the specifications with very little or no inspection required.

JIT was developed and perfected by *Taiichi Ohno (also known for the Eight Wastes)*[51] of Toyota Corporation during 1960 - 70s to meet the fast changing consumer demands with minimum delays.

JIT effective application cannot be independent of other key components of a Lean Manufacturing system or it can "*end up with the opposite of the desired result*".

Lean Manufacturing (*Lean*)[52] is a production practice that considers the expenditure of resources for any goal other than the creation of value for the end customer to be wasteful, and thus a target for elimination.

Working from the perspective of the customer who consumes a product or service, "*value*"[53] is defined as any action or process that a customer would be willing to pay for. Lean is centered on preserving the *value* with less work.

JIT and Lean are tools can require a significant investment of time and money in order to reap the rewards; *but the efforts are well worth it.* Companies utilizing these tools boast of *20% - 30%* overall savings in supplier management and procurement efforts.

This is quite impressive and if done correctly, this philosophy can significantly improve the company bottom line.

Regardless of the methods chosen for the company, the most important factors remain: *people and communication.*

Remember, supplier management is a *relationship* and if approached from that aspect, the right foundation will be set for future success.

Chapter 10

There is a business adage that states: "*If you can't measure it, you can't manage it*". World-class companies measure what is critical to their success.

These measurements, often referred to as *Key Performance Indicators* (KPIs), form a metric that can be measured, tracked and targeted with a strategic focus.

Without KPIs and without strong leadership, a business is blind and unaware of where it *is* or where it is *headed*, and the company performance suffers severely. In addition, a job that lacks clearly defined measurements usually lacks clearly defined goals for its employees. In essence, you are paying an employee to "*not perform their job*". Would you hire a contractor to build an additional room onto your home without giving them the *dimensions and measurements of the intended structure*? Of course not.

If the difference between the *Management* and *Leadership* philosophies are difficult to understand for some, the idea of the *Employee Empowerment* tool can truly baffle those managers who really should be using it on a daily basis. It is illusive and difficult to quantify but the rewards for proper application can be monumental and overall astronomical. "*Communication and information are the lifeblood of empowerment.*"[54]

Quality Management Integration

With its roots in the human relations movement of the 1950 – 60s, the concept of *Employee Empowerment* has a lengthy history related to various theories and techniques designed to democratize the workplace (*Pett and Miller 1994, 153*).

Employee Empowerment[55] is a process of enabling and supporting an organization's human resources to make high-quality, efficient, and effective decisions leading to continuous quality improvement.[56]

Essentially, *Employee Empowerment* involves transferring decision-making authority and responsibility from management to the employees.

Empowering employees is a *top - down* change that must begin with upper management support[57]. The *Manager* must be empowered by upper management and in turn, empowerment can be used by the "*Manager - turned - Leader*".

A *win-win* attitude replaces the old *win-lose* attitude for those leaders who are successful implementing empowerment initiatives. As the benefits of empowering employees become apparent, the properly trained leader will become a strong proponent of empowerment and recognize the value inherent in taking advantage of everyone's experience and creativity.

If one accepts the premise that empowered employees are more satisfied with their jobs, and the satisfied employees result in satisfied customers, then transitional logic dictates that leaders will seek every empowerment opportunity available in an effort to grow the business and increase revenues.

The concept of empowerment is broader than the related concepts of delegation, job enrichment, and participatory management; yet, narrower than the concept of *Employee Involvement*[58]. Hence, *Employee Empowerment* departs from these earlier approaches to workplace democratization both in terms of scope and role of employee participation in the (*quality improvement*) decision-making process.

Motivated, empowered employees are more productive. *"The difference between mediocre and excellent [employees] depends on how the employee is managed."*[59]

Communication is paramount in employee empowerment cultures. In the absence of communication, employees do not know the ramifications of their actions and therefore *are not responsible*. The objectives and direction of the company should be verbally communicated to employees as often as possible.

There are three basic steps in the process of empowerment:

Quality Management Integration

1) The creation of employee responsibility;
2) Authority; and
3) Accountability.[60]

Responsibility is *"The obligation by the subordinate to the supervisor to carry out assigned duties, as agreed."*

Authority is *"Delegated by the superior to provide the subordinate with sufficient rights of command to carry out the assigned tasks."*

Accountability is *"The requirement for a subordinate to answer to the supervisor for results accomplished in the performance of any assigned duties".*[61]

Empowerment also differs from *Job Enrichment,* a job design strategy aiming to expand an employee's job depth; that is, the extent of planning and evaluating duties performed by the employee rather than the supervisor.[62]

Based on the *Two-factor Theory,* all jobs in a work environment may be thought of as having two dimensions: *content and context.* Job content represents *"The tasks and procedures necessary for carrying out a particular job".*[63]

Job context is much broader. Essentially, it is *"The reason the organization needs that job done and includes both how it fits into the overall organizational mission, goals, and objectives and the organizational setting within which that job is done*".[64]

Empowerment differs from *Job Enrichment* in that empowerment focuses primarily on job context rather than job content. Under an empowerment strategy, employees as *process-owners*, experience a high degree of interaction and interdependency.[65]

For this reason, they should be *authorized* to make decisions about job context, as well as job content, in order to materialize the quality improvement goals they are both *individually* and *collectively* responsible and accountable for in daily job performance.

Empowerment is different from *Employee Involvement* in that it constitutes one form of involvement in *Total Quality Management* (*TQM*)[66], alongside the alternative intrinsic and / or extrinsic employee involvement practices.[67]

Specifically, empowerment departs from involvement in terms of the increased scope of employee decision-making responsibility and authority provided under the former.

The value of providing a compelling vision of an empowered workplace should not be underestimated. Because empowerment is often poorly understood, and usually has not been experienced by employees, it is the vision of what is possible that brings their commitment to it.

Vision is perhaps the most visible component of organizational culture; it is through the vision of what is possible that leaders can inspire employees to apply their skills, knowledge, and creativity towards its achievement. Whatever the mind of man can conceive and believe, *it can achieve*.

In order to implement employee empowerment, the employees must be competent. Competency goes beyond developing job-task specific knowledge.

"Employees must be properly trained. It does not make sense to empower employees to do things such as make decisions or approve / initiate action if they are not properly trained."[68] Employees training should include the reason their job is needed and how it fits together with other company processes. Employees should also know who their customers are: *internal and external*.

An effective way to use the empowerment resources is through a *quality circle*[69]. This is a self-directed team that works together to achieve a common objective.

The benefits of empowerment will become clear as the team uses their intellect, imagination, innovation and problem-solving skills to find solutions to sometimes difficult company problems such as *Theory of Constraints (TOC)*[70] issues.

Chapter 11

A *process* is a collection of interrelated work tasks initiated in response to an event that achieves a specific result for the customer of the process.

A process must deliver a specific *result*. This result must be individually identifiable and countable. A good process name clearly indicates the result or end state of the process.

Adding more specific detail to that general definition:

1. The customer of the process can be:
   - A customer receives the result or is the beneficiary;
   - The customer can be a person, stakeholder and / or organization;
   - Customer can pass judgment on the result and process;
   - Customer point of view identifies the process accurately .

2. Initiated in response to a specific event:
   - Multiple events may initiate a process;
   - Having an event and a result allows the tracing of the sequence of tasks.

3. Work tasks:
   - A collection of actions, activities or steps that make up a business process;
   - A step in the initial workflow will probably be divided into more detailed steps later in

the process;

4. A collection of interrelated steps:
    - The process steps must relate to each other through sequence and flow. The completion of one step leads to (*flows into*) the initiation of the next step;
    - Also interrelated with the same work item;
    - Steps related by being traceable back to the same initiation event .

The scientific method evolved over time, with some of history's greatest and most influential minds adding to and refining the process. The Greeks were the first Western civilization to adopt observation and measurement as part of learning about the world, there was not enough structure to call it the scientific method. The Egyptians used processes in constructing the Great Pyramids.

The development of a *scientific process* resembling the modern method was developed by Muslim scholars, during the Golden age of Islam, and refined by the enlightenment scientist-philosophers. The scientific management/approach[71] evolved from the scientific process.

Business Process Management is a way of looking at and controlling the processes that are present in an organization. It is an effective methodology to use in times of crisis to make certain that the processes are efficient and effective, as this will result in a better and more cost efficient organization.

This method is often used together with *Business Process Reengineering (BPR)*[72] to monitor and adjust the output of a process to optimize the process and minimize waste.

Closely related is the *Deming Cycle*[73], which uses logical steps to plan, do, check (*study*) and act the process. It must be used as an iterative methodology in order to continuously evaluate the process and make changes as needed to ensure the desired output.

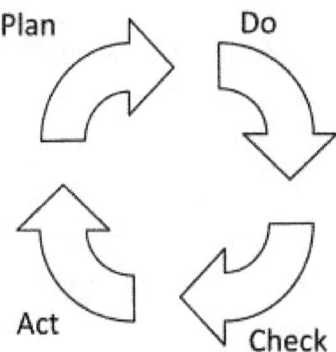

Figure 18: The Deming Cycle

During the planning phase, all stakeholders should be consulted and all resources exploited to ensure the creation of a viable and validated process model.

The first question is: *"What is the purpose of the process?"* This simple question is the rationale behind the justification to expend resources to create the new process.

If you do not know the answer to this fundamental question, it is probably a good time to ask, *"What is the business need of the new process?"* The simplest method is to *begin at the end* and ask *"What will this process produce?"*

Every process has an *input (the products, services and material obtained from suppliers to produce the outputs delivered to customers)* and an *output (products, materials, services or information provided to customers (internal or external), from a process)*. The output is the *reason* for the process' existence.

A process may function in isolation to produce an output which is directly consumed by a customer or it may work in harmony with other processes to produce outputs need for a more overarching process. Either way, the process must be created and managed.

Once the process has been planned, it should be drafted into a *flowchart*[74]. This tool enables the visual evaluation of the process from the functional perspective.

If the process is one of several processes of a system, the flowchart will allow the flow to be analyzed and the steps to be viewed through the eyes of the person(s) performing the steps of the process. The flowchart is a good analytical tool for identifying excessive steps and actions that are not value-added.

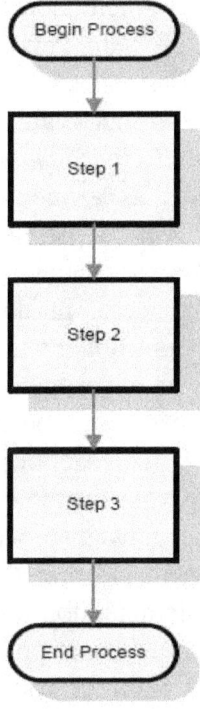

Figure 19: Process Flowchart

Figure 19 illustrates how the flowchart can help with visualizing process steps as well as the overall structure and construction of the process. It can be evaluated at this point and a determination made to add, delete, or in any other manner, amend the steps to produce the desired results.

The Deming Cycle methodology then focuses attention on the *"Do"* step. This step is where the process is tested in a small controlled environment so the results can be further evaluated. The key is to remember *"controlled environment"*. While this is considered *laboratory experimentation,* the benefit is this will protect your existing *"live"* processes just in case an unforeseen flaw exists within the steps of the new process. In doing so, you can protect your customers from any negative results.

Also remember it *is* an experiment and the time will come when the process must be implemented into the system and negative reactions and / or results may occur allowing your customers to *"experience the effects"* of the flawed process. Use of the controlled environment to *brainstorm*[75] can identify potential future implementation problems. This will lower the chances of a catastrophic failure once the process *"goes live"* within the system structure.

A process should be designed with controls in mind which will assist the users to remain within the boundaries of the process scope.

Controls / decision points may also be designed as *checkpoints* or *process gates* that allow users to quickly determine if the process is functioning properly. These may or may not be auditable points but the idea is not to require constant audits / inspections of the process. If proper controls are built in, a certain level of confidence will be provided as a virtual process byproduct.

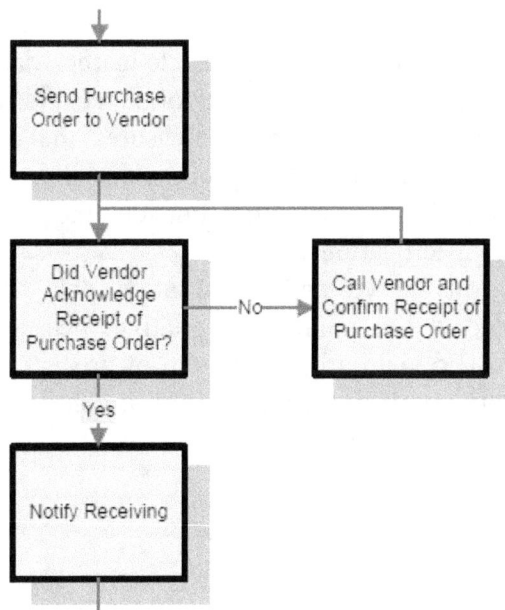

Figure 20: Process Decision / Control Point

If the user tests the process during application and

finds the process too cumbersome or confusing, this issue must be addressed immediately with the input of all stakeholders. If left unaddressed, this issue may result in a total failure of the process and all processes that are overall system interdependent.

Human nature *will* surely prevail and the user feedback is valuable and of utmost importance at this stage of process development. There is no limit to how a process can be designed, but simple will serve most users very well.

These points can also be used to gather data and metrics to indicate process performance and trends. *Statistical Process Control* measures the process and can quickly identify a process that is going out of control[76] or deviating outside the required process specifications.

While qualitative and quantitative data can be collected at the points, quantitative measures are desired and yield a better picture of process trends. The data points can be plotted on a control chart[77] for easy dissemination.

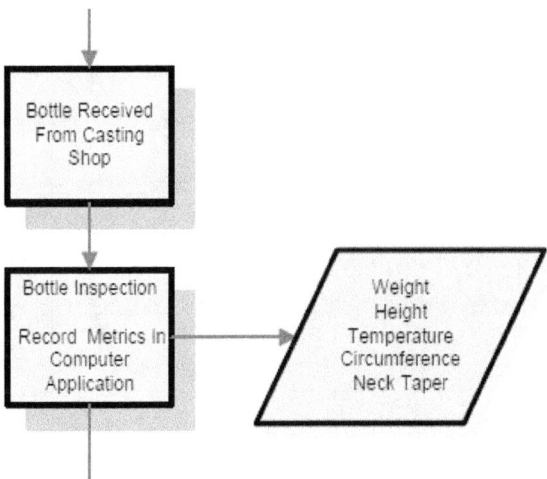

Figure 21: Process Data Collection Point

The next element in the process creation is *"check"* or *"study"*. In this step, the results of the test are reviewed and analyzed according to the expected results. Any inconsistencies or anomalies are identified and considered for disposition. Any lessons learned are also documented and the outcomes evaluated.

Amendments to the process may take place at this point to correct or prevent any foreseen problems. This is a crucial step and should be viewed as the

Page 111

opportunity not only to correct problems, but also to look for opportunities for improvement.

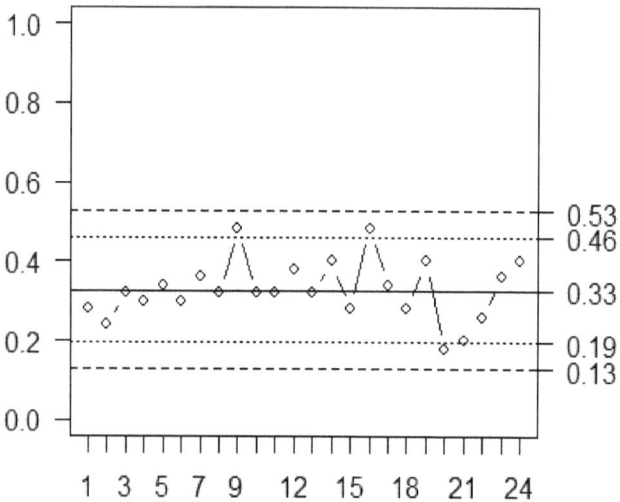

Figure 22: Control Chart

If control charts (*Figure 22*) are used to evaluate the process, key data points can help zero in on the areas that are unstable or simply out of control. There are a few general rules to keep in mind when reviewing a process control chart.

Any single data point falling above the +3σ limit

Two consecutive points falling above the +2σ limit
Three consecutive points falling above the +1σ limit
Seven consecutive points falling above the centerline
Ten consecutive points falling below the centerline
Six consecutive points falling below the -1σ limit
Four consecutive points falling below the -2σ limit

Figure 23: R chart rules – Process out of control

Standard Deviation represents + / - 3σ from the mean on a control chart. The significance of this is the normal distribution of data points will usually fall within this + / - 3σ area. Any points that fall outside this normal area should be investigated immediately.

If the number or orientation of the data points matches the chart rules, the process is *not in control*. Correction and / or prevention measures should be initiated.

$$\sigma = \sqrt{\frac{\sum (x - \overline{x})^2}{n}}$$

σ = **lower case sigma**
$\sum$ = **capital sigma**
$\overline{x}$ = **x bar**

Figure 24: Standard Deviation Formula

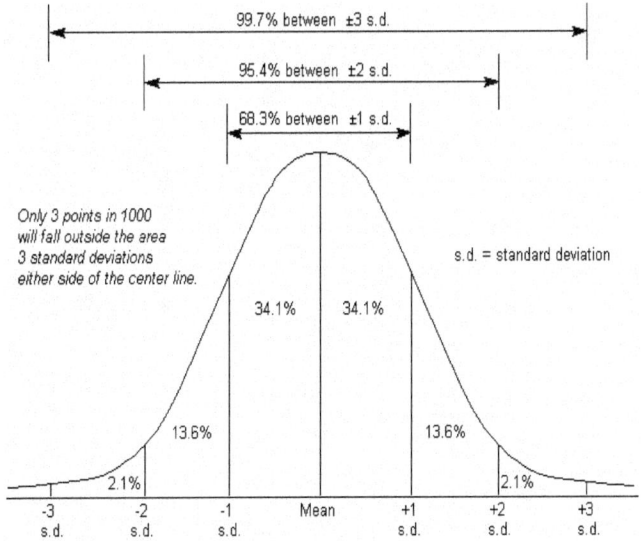

Figure 25: Standard Deviation / Normal Distribution

Quality Management Integration

Chapter 12

The employees are the most important resource of any company. It is important to build a cooperative and open environment between employees and management in order to realize the true talents and expertise of the employees. If management loses sight of this critical requirement, management may soon be looking for other jobs.

In *Quality Management*, most of the work is actually performed by teams. Examples include *Six Sigma*[78] process teams, auditor teams, project teams, steering committees and other countless team applications. Since teams will be an integral and essential part of the organization as a whole, the team concepts merit further exploration in this chapter.

Teams are valuable when forming new process plans, reviewing progression and solving problems that may have risen from process flaws. Bringing together those individuals having subject matter knowledge can reap rewards for any department.

Brainstorming is an important tool the team can use to identify root causes of problems. This consensus approach taps the knowledge and experience of the team members and can result in cost savings due to process optimization initiatives.

Understanding team dynamics, limitations and strengths is important for management. People who are told to work together *do not* constitute a team. There is a vast difference in the team and a group. The two are not synonymous. The results from each differ greatly.

Team selection is an important step and should not be taken lightly. The members selected will be the ones yielding the results and the right people need to be selected for the job. Selecting someone for any other reason that the value they will bring to the team is a major mistake. The members do not necessarily need to have worked on teams in the past. Each team will go through a process together that will educate and prepare them for the work ahead.

Equally important is the person who will lead or facilitate (*facilitator*)[79] the team as the work is accomplished. This person will play a vital role in keeping the objective of the team in sight and preventing the team from straying outside the scope (*scope creep*) of what is expected from the team.

Once the team members are chosen, the *problem statement*[80] must be clearly explained to the new team. Also, *ground rules* should be introduced. These rules will govern the actions and accepted conduct of the individuals.

Teams will go through stages of development as the members become familiar with one another and begin to work on the problem.

Quality Management Integration

Bruce Tuckman defined these stages and most teams will follow the process to some degree. Perhaps all steps will be easily defined but some may occur concurrently. These stages are: *Forming, Storming, Norming, Performing and Adjourning.*

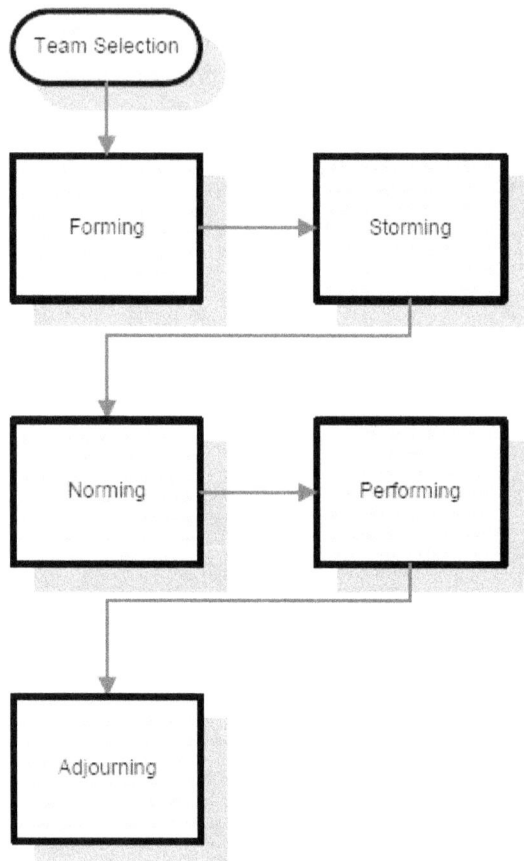

Figure 26: The Five Team Stages

The team dynamic changes as the team moves from stage to stage. During the *forming* stage, serious issues and feelings are avoided, and people focus on being busy with routines, such as team organization, roles, etc.

Individuals are also gathering information and impressions about each other, about the scope of the task and how to approach it. This is a comfortable stage, but due to the avoidance of conflict and threat, very little is actually accomplished.

The *storming* stage can be the most difficult for team members. Team members open up to each other and confront each other's ideas and perspectives. In some cases *storming* can be resolved quickly.

Tolerance of each team member and their differences needs to be emphasized. Without tolerance and patience, the team will fail. This phase can become destructive to the team and will lower motivation if allowed to get out of control. Some teams will never develop past this stage.

During the *norming* stage, the team begins to relax and rely on one another to begin moving in the right direction together. Differences are put aside and outstanding issues are resolved. The team focuses on the problem instead of each other. *Storming* can still occur at this point but if the process is moving in the right direction, the disruption is usually minimal.

The *performing* stage is when the actual work is accomplished. The team members are now competent, autonomous and able to handle the decision-making process without supervision. Dissent views are expected and allowed as long as it is channeled through means acceptable to the team.

A change in *leadership* may cause the team to revert to *storming* as the new people challenge the existing norms and dynamics of the team.

Even the most high-performing teams will revert to earlier stages in certain circumstances. Many long-standing teams go through these cycles many times, as they react to changing circumstances. As the project comes to a close or the problem has been solved, the team may disband. This is sometimes referred to as *adjourning* or *mourning*.

Models such as the *Types of Work Wheel* give a reliable and valid way of measuring and managing team performance, by generating qualitative and quantitative feedback data both from team members and outsiders. Problems can be diagnosed or even predicted before they happen.

In managing team performance, clever work teams will use this information to bypass the *storming* stage and move quickly to the *norming* stage by generating ground rules which will prevent major problems from occurring. The team can then accelerate its progress to the *performing* stage.

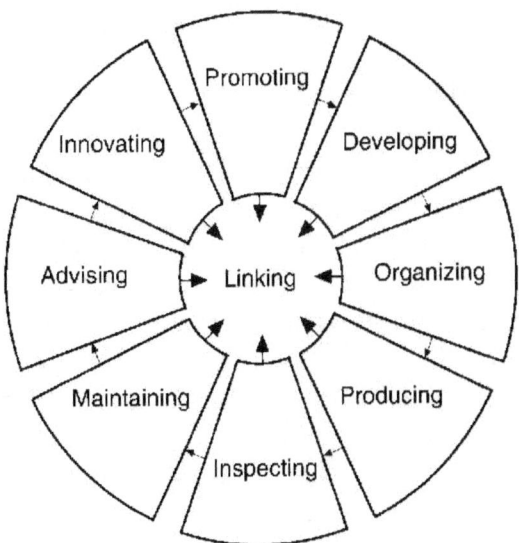

Figure 27: Types of Work Wheel

Team Performance Management is focused directly on the achievement of the team's key business objectives. It bridges the gap between the team building 'enablers' and business performance results. It removes the reliance on *"faith"*: *the need to believe that team building works before investing in it.* This establishes a direct connection between collective behaviors and team performance.

Team performance metrics are defined measures of progress associated with a team's performance. As opposed to measuring an individual employee, this metric measures the team's overall performance as a whole.

Team performance metrics may also be used to measure the progress and success of similar teams in an effort to create a benchmarking system. This kind of metric can indicate to management areas for improvement, as well as strengths of the team.

It can also show when individuals are not performing up to standards and are negatively affecting the overall performance of the team. Teams may also face some common problems through the various stages.

- *Absence of team identity*: Members may not feel mutually accountable to one another for the team's objectives. There may be a lack of commitment and effort, conflict between team goals and members' personal goals, or poor collaboration.
- *Difficulty making decisions:* Team members may be rigidly adhering to their positions during decision making or making repeated arguments rather than introducing new information.
- *Poor communication:* Team members may interrupt or talk over one another. There may

be consistent silence from some members during meetings, allusions to problems but failure to formally address them, or false consensus (everyone nods in agreement without truly agreeing).

- *Inability to resolve conflicts:* Conflicts cannot be resolved when there are heightened tensions and team members make personal attacks or aggressive gestures.
- *Lack of participation:* Team members fail to complete assignments. There may be poor attendance at team meetings or low energy during meetings.
- *Lack of creativity:* The team is unable to generate fresh ideas and perspectives and does not turn unexpected events into opportunities.
- *Groupthink:* The team is unwilling or unable to consider alternative ideas or approaches. There is a lack of critical thinking and debate over ideas. This often happens when the team overemphasizes team agreement and unity.
- *Ineffective leadership:* Leaders can fail teams by not defining a compelling vision for the team, not delegating, or not representing multiple constituencies.

In terms of corporate development, team-building exercises are important not for the immediate experience of the activities performed by the team, but also for the group skills, communication and bonding that result.

The main goals of team building are to improve productivity and motivation. Taking employees out of the office helps groups break down political and personal barriers, eliminate distractions, and have fun.

The benefits of team-building programs are so significant that many corporations have incorporated teambuilding strategies into their standard training curriculum. Activities in these programs are designed to motivate people to pool their talents and perform at their best individually and as team players.

Team members discover that diversity is their greatest asset and trust, cooperation and effective communication are the key to a team's success. Structured activities not only encourage individual development, but also bring all members together for a common cause. Team building programs provide upbeat and powerful team experiences allowing companies to compete effectively by enabling staff at all corporate levels to work as true team players.

Team building generally sits within the theory and

practice of organizational development. The related field of team management refers to techniques, processes and tools for organizing and coordinating a team towards a common goal as well as the inhibitors to teamwork and ways to remove, mitigate or overcome them.

In the organizational development context, a team may embark on a process of self - assessment to gauge its effectiveness and improve its performance.

To assess itself, a team seeks feedback from group members to find out both its current strengths and weakness. To improve its current performance, feedback from the team assessment can be used to identify gaps between the desired state and the current state, and to design a gap-closure strategy.

Virtual teams are a great way to enable teamwork in situations where people are not sitting in the same physical office at the same time. Virtual teams are governed essentially the by same fundamental principles as traditional teams. Yet, there is one critical difference.

This difference is the way the team members communicate. Instead of using the full spectrum and dynamics of in-office face-to-face exchange, they now rely on special communication channels enabled by modern technologies.

Due to more limited communication channels, the

success and effectiveness of virtual teams is much more sensitive to the type of project the group works on, what people are selected, and how the team is managed.

Managers of virtual teams also must pay much more attentions to maintaining clear goals, performance standards, and communication rules. People have varying assumptions of what to expect from each other.

To avoid build-ups of misunderstandings, in a virtual organization it is critical to replace those implicit assumptions with clear rules and protocols that everyone understands and agrees upon, especially for communication.

One of the biggest challenges of virtual teams is building and maintaining trust between the team members. Trust is critical for unblocking communication between members and sustaining motivation of each person involved. The issue of trust needs special attention at any stage of team existence.

As with any type of team in any company, there is only one way that a team can be beneficial to the company and that is support. Most teams fail within the first week of inception due to lack of support.

A process improvement team can be extremely beneficial if it has the support of all the departments and the head of the company.

Most improvement teams follow principles either learned through training within the company or by the company hiring a specialist who has been trained for many months or years in *Lean* principles and *Six Sigma* methods.

Once an area for improvement is chosen the scope and purpose of the project is chosen and documented to keep from deviating, the actual improvement project is defined and documented and as many of the root causes of inefficiencies or problems should be identified and documented.

Quality Management Integration

List of Figures and Tables

Quality Management Integration

Quality Management Integration

[1] Glossary of Terms

[2] *SPC Definition: The application of statistical techniques to control a process; often used interchangeably with the term "statistical quality control."*

[3] *The Human Side of Enterprise* (1960) - Douglas McGregor

[4] Dictionary.com

[5] http://giftedkids.about.com/od/glossary/g/extrinsic.htm

[6] *Strategic Planning: The process an organization uses to envision its future and develop the appropriate strategies, goals, objectives and action plans.*

[7] *Definition: The process of considering the strategic position of the organization and using input from all sources, planning the future position of the company.*

[8] *Goetsch*, 2000

[9] *Groupthink: A situation in which critical information is withheld from the team because individual members censor or restrain themselves, either because they believe their concerns are not worth discussing or because they are afraid of confrontation.*

[10] *Brainstorming: A technique teams use to generate*

Quality Management Integration

ideas on a particular subject. Each person on the team is asked to think creatively and write down as many ideas as possible. The ideas are not discussed or reviewed until after the brainstorming session.

[11] *Risk Management: Using managerial resources to integrate risk identification, risk assessment, risk prioritization, development of risk handling strategies and mitigation of risk to acceptable levels.*

[12] *Porter*, 1979

[13] *Corrective action: A solution meant to reduce or eliminate an identified problem.*

[14] *Preventive action: Action taken to remove or improve a process to prevent potential future occurrences of a nonconformance.*

[15] *Root cause: A factor that caused a nonconformance and should be permanently eliminated through process improvement.*

[16] *Auditee: Entity or person(s) being audited.*
http://www.businessdictionary.com/definition/auditee.html

[17] *Finding: A conclusion of importance based upon observations.*

[18] *Nonconformity: The non-fulfillment of a specified*

requirement. Also see "blemish," "defect" and "imperfection."

[19] *Process to verify corrective and / or preventative action plans and implementations.*

[20] *Six Sigma: A method that provides organizations tools to improve the capability of their business processes. This increase in performance and decrease in process variation lead to defect reduction and improvement in profits, employee morale and quality of products or services. Six Sigma quality is a term generally used to indicate a process is well controlled (±6 s from the centerline in a control chart).*

[21] *Quality Management System*

[22] *Metric: A standard for measurement.*

[23] *Pareto chart: A graphical tool for ranking causes from most significant to least significant. It is based on the Pareto principle, which was first defined by Joseph M. Juran in 1950. The principle, named after 19th century economist Vilfredo Pareto, suggests most effects come from relatively few causes; that is, 80% of the effects come from 20% of the possible causes. One of the "seven tools of quality".*

[24] *Joseph M. Juran*

[25] *Histogram: A graphic summary of variation in a set of*

data. The pictorial nature of a histogram lets people see patterns that are difficult to detect in a simple table of numbers. One of the "seven tools of quality".

[26] *Ogive: A distribution curve in which the frequencies are cumulative.*

[27] *Stem and Leaf: a method for showing the frequency with which certain classes of values occur.*

[28] *Control chart: A chart with upper and lower control limits on which values of some statistical measure for a series of samples or subgroups are plotted. The chart frequently shows a central line to help detect a trend of plotted values toward either control limit.*

[29] *Control limits: The natural boundaries of a process within specified confidence levels, expressed as the upper control limit (UCL) and the lower control limit (LCL).*

[30] *Upper control limit (UCL): Control limit for points above the central line in a control chart. Lower control limit (LCL): Control limit for points below the central line in a control chart.*

[31] *Truncated: Having the apex, vertex, or end cut off by a plane: a truncated cone or pyramid.*

[32] *Poisson distribution: A discrete probability distribution that expresses the probability of a number*

of events occurring in a fixed time period if these events occur with a known average rate, and are independent of the time since the last event.

[33] *Scatter diagram: A graphical technique to analyze the relationship between two variables. Two sets of data are plotted on a graph, with the y-axis being used for the variable to be predicted and the x-axis being used for the variable to make the prediction. The graph will show possible relationships (although two variables might appear to be related, they might not be; those who know most about the variables must make that evaluation). One of the "seven tools of quality".*

[34] *Cause and effect diagram: A tool for analyzing process dispersion. It is also referred to as the "Ishikawa diagram," because Kaoru Ishikawa developed it, and the "fishbone diagram," because the complete diagram resembles a fish skeleton. The diagram illustrates the main causes and sub-causes leading to an effect (symptom). The cause and effect diagram is one of the "seven tools of quality".*

[35] *Flowchart: A graphical representation of the steps in a process. Flowcharts are drawn to better understand processes. One of the "seven tools of quality".*

[36] *Cost of quality (COQ): Another term for COPQ. It is considered by some to be synonymous with COPQ but is considered by others to be unique. While the two concepts emphasize the same ideas, some disagree as to*

which concept came first and which categories are included in each.

[37] *Muda: Japanese for waste; any activity that consumes resources but creates no value for the customer.*

[38] *Activity Based Costing: An accounting system that assigns costs to a product based on the amount of resources used to design, order or make it.*

[39] *Do It Right The First Time*

[40] *A source of materials, service or information input provided to a process.*

[41] *Confidence a supplier's product or service will fulfill its customers' needs. This confidence is achieved by creating a relationship between the customer and supplier that ensures the product will be fit for use with minimal corrective action and inspection. According to Joseph M. Juran, nine primary activities are needed: 1. define product and program quality requirements; 2. evaluate alternative suppliers; 3. select suppliers; 4. conduct joint quality planning; 5. cooperate with the supplier during the execution of the contract; 6. obtain proof of conformance to requirements; 7. certify qualified suppliers; 8. conduct quality improvement programs as required; 9. create and use supplier quality ratings.*

[42] The Association for Operations Management

Quality Management Integration

[43] *The series of suppliers to a given process.*

[44] Mentzer et. al., 2001

[45] *Highest current performance level in an industry, used as a standard or benchmark to be equaled or exceeded.*

[46] *Released for the first time in October 1994, an economic indicator and cross industry measure of the satisfaction of U.S. household customers with the quality of the goods and services available to them. This includes goods and services produced in the United States and imports from foreign firms that have substantial market shares or dollar sales. ASQ is a founding sponsor of the ACSI, along with the University of Michigan Business School and the CFI Group.*

[47] *The overseeing of process instances to ensure their quality and timeliness; can also include proactive and reactive actions to ensure a good result.*

[48] *Supply Chain Management: Processes, Partnerships, Performance*, 3rd edition, 2008

[49] *The rate of customer demand, takt time is calculated by dividing production time by the quantity of product the customer requires in that time. Takt is the heartbeat of a lean manufacturing system.*

[50] *Japanese for waste; any activity that consumes resources but creates no value for the customer.*

[51] *Taiichi Ohno originally enumerated seven wastes (muda) and later added underutilized people as the eighth waste commonly found in physical production. The eight are: 1. overproduction ahead of demand; 2. waiting for the next process, worker, material or equipment; 3. unnecessary transport of materials (for example, between functional areas of facilities, or to or from a stockroom or warehouse); 4. over-processing of parts due to poor tool and product design; 5. inventories more than the absolute minimum; 6. unnecessary movement by employees during the course of their work (such as to look for parts, tools, prints or help); 7. production of defective parts; 8. under-utilization of employees' brainpower, skills, experience and talents.*

[52] *An initiative focused on eliminating all waste in manufacturing processes. Principles of lean manufacturing include zero waiting time, zero inventory, scheduling (internal customer pull instead of push system), batch to flow (cut batch sizes), line balancing and cutting actual process times. The production systems are characterized by optimum automation, just-in-time supplier delivery disciplines, quick changeover times, high levels of quality and continuous improvement.*

[53] *A term used to describe activities that transform input into a customer (internal or external) usable output.*

Quality Management Integration

[54] Ginnodo, 1997, p. 12

[55] *A condition in which employees have the authority to make decisions and take action in their work areas without prior approval. For example, an operator can stop a production process if he or she detects a problem, or a customer service representative can send out a replacement product if a customer calls with a problem.*

[56] Dimitriades 1999b, 17

[57] *Participation of the highest level officials in their organization's quality improvement efforts. Their participation includes establishing and serving on a quality committee, establishing quality policies and goals, deploying those goals to lower levels of the organization, providing the resources and training lower levels need to achieve the goals, participating in quality improvement teams, reviewing progress organization-wide, recognizing those who have performed well and revising the current reward system to reflect the importance of achieving the quality goals.*

[58] *An organizational practice whereby employees regularly participate in making decisions on how their work areas operate, including suggestions for improvement, planning, goal setting and monitoring performance.*

[59] Blanchard, et. al. 2003

Quality Management Integration

[60] Newman, Warren, and Schne 1982, 221

[61] Schermerhorn 1986, 177

[62] Schermerhorn 1986, 222

[63] Ford and Fottler 1995

[64] Ford and Fottler 1995

[65] Caudron 1995

[66] *A term first used to describe a management approach to quality improvement. Since then, TQM has taken on many meanings. Simply put, it is a management approach to long-term success through customer satisfaction. TQM is based on all members of an organization participating in improving processes, products, services and the culture in which they work. The methods for implementing this approach are found in the teachings of such quality leaders as Philip B. Crosby, W. Edwards Deming, Armand V. Feigenbaum, Kaoru Ishikawa and Joseph M. Juran.*

[67] Dimitriades 1999a, 235–236

[68] Gandz, 1990, p. 76 Byham 1997

[69] *A quality improvement or self-improvement study group composed of a small number of employees (10 or*

fewer) and their supervisor. Quality circles originated in Japan, where they are called quality control circles.

[70] *A lean management philosophy that stresses removal of constraints to increase throughput while decreasing inventory and operating expenses. TOC's set of tools examines the entire system for continuous improvement. The current reality tree, conflict resolution diagram, future reality tree, prerequisite tree and transition tree are the five tools used in TOC's ongoing improvement process. Also called constraints management.*

[71] *A term referring to the intent to find and use the best way to perform tasks to improve quality, productivity and efficiency.*

[72] *The concentration on improving business processes to deliver outputs that will achieve results meeting the firm's objectives, priorities and mission.*

[73] *Another term for the plan-do-study-act cycle. Walter Shewhart created it (calling it the plan-do-check-act cycle), but W. Edwards Deming popularized it, calling it plan-do-study-act.*

[74] *A graphical representation of the steps in a process. Flowcharts are drawn to better understand processes.*

[75] *A technique teams use to generate ideas on a particular subject. Each person on the team is asked to think creatively and write down as many ideas as*

possible. The ideas are not discussed or reviewed until after the brainstorming session.

[76] *A process in which the statistical measure being evaluated is not in a state of statistical control. In other words, the variations among the observed sampling results can be attributed to a constant system of chance causes.*

[77] *A chart with upper and lower control limits on which values of some statistical measure for a series of samples or subgroups are plotted. The chart frequently shows a central line to help detect a trend of plotted values toward either control limit.*

[78] *A method that provides organizations tools to improve the capability of their business processes. This increase in performance and decrease in process variation lead to defect reduction and improvement in profits, employee morale and quality of products or services. Six Sigma quality is a term generally used to indicate a process is well controlled (±6 from the centerline in a control chart).*

[79] *A specifically trained person who functions as a teacher, coach and moderator for a group, team or organization.*

[80] *The description of the issue or the problem which needs to be addressed by a problem solving team.*

Quality Management Integration

About The Author

Warren Alford grew up in Louisiana. He learned to fly at age seventeen and after obtaining his Commercial pilot's license, flew worldwide as an airline pilot in a career spanning over twenty years.

Warren holds the following professional quality memberships and certification:

ASQ Certified Manager of Quality / Organizational Excellence (CMQ / OE)
ASQ Certified Quality Auditor (CQA)
PMI Project Management Professional (PMP)

Warren currently manages the local QMS for a global simulation company. In this position, he is responsible for planning / scheduling the Quality Monitoring Program, including auditor selection and training, performing audits and keeping upper

Quality Management Integration

management informed of the QMS status.

His paperback books, audio books and albums are distributed world - wide by Speckbohne Media.

For more information please visit
www.WarrenAlford.com.

Other Offerings From Speckbohne Media

Quality Management Integration
(Paperback Book)
ISBN-10: 145385472X

Quality Management Integration
(Kindle Book)
ASIN: B0044XUWJS

Hope For The Righteous
(Paperback Book)
ISBN-10: 1452803315

Hope For The Righteous
(Kindle Book)
ASIN: B003IHW1GS

Relipocrisy
(Paperback Book)
ISBN-10: 1450547796

Relipocrisy
(Kindle Book)
ASIN: B0036FU1A0.

www.ingramcontent.com/pod-product-compliance
Lightning Source LLC
Chambersburg PA
CBHW051531170526
45165CB00002B/687